Workshop on Materials and Engineering in Aeronautics (MEA2019)

IOP Conference Series: Materials Science and Engineering
Volume 714

Moscow, Russia
16 - 17 October 2019

Editors:

Evgeny V. Nikulchev
Oleg N. Korsun

ISBN: 978-1-7138-2573-9
ISSN: 1757-8981

Printed from e-media with permission by:

Curran Associates, Inc.
57 Morehouse Lane
Red Hook, NY 12571

Some format issues inherent in the e-media version may also appear in this print version.

Institute of Physics
Dirac House, Temple Back
Bristol BS1 6BE UK

Phone: 44 1 17 929 7481
Fax: 44 1 17 920 0979

techtracking@iop.org

Additional copies of this publication are available from:

Curran Associates, Inc.
57 Morehouse Lane
Red Hook, NY 12571 USA
Phone: 845-758-0400
Fax: 845-758-2633
Email: curran@proceedings.com
Web: www.proceedings.com

TABLE OF CONTENTS

IOP Conf. Series: Materials Science and Engineering **714** (2020) 011001 doi:10.1088/1757-899X/714/1/011001

PREFACE

2019 Workshop on Materials and Engineering in Aeronautics (MEA2019) was held 16-17 October 2019 in Moscow Aviation Institute (National Research University).

This Workshop was held 15 times from 2001 to 2018. Each was attended by over 20 participants who submitted their reports for publishing. The participants represented mostly the leading technical universities of Russia. The Scientific Committee hopes that the international involvement would increase.

The (MEA2019) Workshop Proceedings include 28 papers but the scope is actually huge. It covers materials & structure development, aircraft and airships design, pilot interface, aircraft parameter identification, system theory, control, aircraft engines technical state monitoring, robots, statistical classification. This year we include more papers dealing with deep learning and recurrent neural networks applications. In the year 2019 the Scientific Committee's confirms its traditional position not to limit the creativity of the participants. As a result all the papers are unlike one another and many of them contain original ideas and approaches. That is why our advice to a reader is not to judge the Proceedings by one or two papers but look through the most part of them.

On behalf of the Workshop Scientific Committee
the scientific editors of MEA2019 Proceedings:

Evgeny V. Nikulchev, Dr., Professor of MIREA – Russian Technological University,
nikulchev@mail.ru

Oleg N. Korsun, Dr., Prof., Head of the laboratory, State Research Institute of Aviation Systems,
marmottoduo@yandex.ru

IOP Conf. Series: Materials Science and Engineering **714** (2020) 011002 doi:10.1088/1757-899X/714/1/011002

Peer review statement

All papers published in this volume of *IOP Conference Series: Materials Science and Engineering* have been peer reviewed through processes administered by the proceedings Editors. Reviews were conducted by expert referees to the professional and scientific standards expected of a proceedings journal published by IOP Publishing.

IOP Conf. Series: Materials Science and Engineering **714** (2020) 012001 doi:10.1088/1757-899X/714/1/012001

The Evolution of Imperative Programming Paradigms as a Search for New Ways to Reduce Code Duplication

T Avacheva[1] and A Prutzkow[2]

[1] Department of Mathematics, Physics, and Medical Computer Sciences, Ryazan State Medical University named after I. P. Pavlov, Ryazan, Russian Federation
[2] Department of Computational and Applied Mathematics, Ryazan State Radio Engineering University named after V. F. Utkin, Ryazan, Russian Federation

E-mail: mail@prutzkow.com

Received xxxxxx
Accepted for publication xxxxxx
Published xxxxxx

Abstract

The cause for imperative programming paradigm shift is the impossibility of developing software systems of a new level of complexity. We consider the evolution of programming paradigms: structured, procedural, and object-oriented. We demonstrate new ways of code duplication reducing have appeared with the shift of paradigm. We conclude the factor of code duplication reducing determines the direction of programming paradigm evolution. We discover the constraints, which were introduced in the paradigms, simplify the development of software systems. We conclude the new constraints allow the development of more complex software systems. The main reason for the code duplication is the low qualification of programmers. Therefore, in the process of learning programming, one should pay attention to code duplication and ways to reduce it.

Keywords: programming paradigms, structured programming, procedural programming, object-oriented programming

1. Introduction

The main objective of the imperative programming paradigm evolution is to decrease expenses of program development as well as maintenance. Programming paradigms shift one another when the old programming paradigm makes developing of a program system of new complexity level impossible. However details are more interesting. Why the programming paradigms are evolving from structured programming, via procedural programming to object-oriented programming? This question inevitably arises when teaching the history of the programming paradigms.

There are ideas of it, such as:

1. "The evolution of modern programming languages is closely coupled with the development (and formalization) of the concept of data type" [1].

2. "This evolution of languages also has been heavily influenced by progress in the theory of computing, which has led to a formal understanding of the semantics of statements, modules, abstract data types, and processes" [2].

3. "Object-oriented programming developed as the dominant programming methodology during the mid-1980s, isn't a revolution in informatics but rather a long evolutionary process aiming to improve the SP [structured programming] efficiency for complex system development" [3].

But there is another idea.

The purpose of writing a program is an algorithmic solution of a problem. However, there may be several solutions. The best solution is chosen in accordance with some criterion related to the time complexity of the problem solution and the complexity of the program structure.

A natural criterion for choosing the structure of a program is the length of its text (number of lines): the shorter the program, the more readable and easier to debug. One of the ways to shorten the text of a program is to use the go to statement. However, small but unstructured programs required a lot of time for maintenance. Therefore, at present, the following is used as a criterion: «program structure should be such as to anticipate its adaptation and modification» [4].

To improve compliance with this criterion, additional program structures were introduced into programming languages, including those associated with the appearance of new programming paradigms. Thus, the driving force behind the evolution of paradigms and programming languages was the need to increase the degree of compliance with the criterion for choosing a program structure. The main way to increase the degree of compliance with the criterion was to reduce code duplication.

2. Code Duplication and Related Problems of Code Reuse and of Redundant Code

Code duplication is the presence in the program text of a fragment that has a functionally equivalent fragment in another part of the program.

Code duplication of program text fragments can be divided into two types:

1. Identical: identical line in text program.
2. Reducible: a fragment can be altered to another one by such as a change of code line sequence, renaming of variable, type or routine names, replacing a control construct by functionally equivalent one, and so forth.

Code duplication complicates a program modification, since it requires making the same changes to its duplicate. The proportion of duplicate lines of text in a program is, according to various estimates, from 5% to 23%, but sometimes it reaches 59% [5–6]. Code duplication is a bad programming practice [7–8]. Along with the code duplication, there are concepts of reusable code and redundant lines of program text.

Code reuse is a practice of reapplying of already developed modules in a new developing program. Also there are reuses of personnel, of specifications, of designs, of «patterns», of specified components, of abstracted modules [9].

Redundant code is a superfluous code, which can be eliminated without program functionality and correctness loss. One kind of redundant code is dead code [7–8].

3. Constraints Imposed by the Programming Paradigms

Each programming paradigm introduces constraints in programming. The constraints put the programmer in some framework that does not allow him to use the capabilities of programming languages, which contradict the paradigm, but reduce the complexity of program development.

4. The Purpose of the Study

The purpose of the study is to analyze the evolution of the paradigms of structured, procedural, and object-oriented programming and to discover ways to reduce code duplication.

Also we will discover the constraints that were introduced in these programming paradigms to simplify the development of software systems.

5. Structured Programming

5.1 The Fundamentals of Structured Programming

The founder of structured programming is Edsger Dijkstra [10, 11]. Structured programming is based on the following fundamentals [4]:

1. The use in the program of control structures of sequence, selection, and repetition, having one entry and one exit. It allows the program to compare the stage of the computational process with a fragment of the program that describes this stage. This mapping is used when debugging a program.
2. Step-wise program composition (top-down development). First, the task is divided into several stages. In the next step, each stage is divided into sub-stages again. This process continues until the stages are described by programming language statements. For a top-down design, only one decision is made at each step (for example, choosing a data structure). The remaining decisions are postponed until development reaches the desired step. Step-wise program composition allowed programs to be made more modular through staged development.
3. Proof of program correctness. «Program testing can be used to show the presence of bugs, but never to show their absence» [12]. To prove the absence of errors in the program, it is necessary to prove the theorem that the program processes any input data into the expected output. The program should be structured, that is, it should include only the structures of sequence, selection and repetition. The proof uses a mathematical apparatus, including mathematical induction.

The first two fundamentals were practically unchanged adopted by following programming paradigms. The third fundamental was never developed into practically applicable [11], therefore it had only theoretical significance.

5.2 Constraints of Structured Programming

Constraints of structured programming are:

1. Use only sequence, selection, and repetition control constructs in programs.
2. Develop a program using step-wise composition.

5.3 Routines in Structured Programming

One of the ways to reduce code duplication is to carry such duplicates into routine. Routines appeared in the 1950s and were used by programmers as a means of reducing the program length, and only in some cases for structured organization [4]. Data is transferred to routines both through parameters and global variables [13]. The use of routines was not a necessary condition for implementing structured programming [11]. The reasons for this attitude of programmers to the routines were as follows:

1. Programs did not reach such a length that would require mandatory division of them into routines.

2. The call of routines slowed down the execution of the programs, since the loading of the routine into memory, its execution, and then the release of this memory, required additional time slow by modern standard processor.

Also the division of the program into modules was optional. One of the criteria for dividing a program into modules is the length of the module [11]. For example, if a module must contain a fragment of program text no more than 50 lines, then the program text with a length of 300 lines will be divided into 6 modules. Such a division is possible through the use of global variables.

The use of structured programming decreases costs in the development of software systems [14, 15]. However, structured programming did not provide a way to reduce code duplication (except cycles), despite the existence of routines in programming languages. Routines were called in a procedural programming paradigm.

6. Procedural Programming

6.1 The Fundamentals of Procedural Programming

The length of program texts grew and it became clear that it was impossible to make a program understandable and easily modifiable by structuring alone. «Structured programming appears to fall apart when applications exceed 100,000 lines or so of code» [16]. Therefore, to reduce code duplication, a concept that existed in structured programming was the use of routine.

Procedural programming appeared when the division of a program into routines in order to avoid code duplication ceased to be a recommendation and became a requirement.

In procedural programming, the concept of levels of abstraction was used [17], which also appeared during the structured programming times. Each level performs a group of related functions [18]. Level functions can only access functions immediately lower level.

We define this paradigm in the follow way.

Procedural programming is a structured programming extended by additional principles:

1. Functions of each level of abstraction are described by routines.

2. Each routine performs one function at its own abstraction level.

3. The routine is single-entry, single-exit control construct, receives the input data through parameters and returns the result through parameters or through the return statement.

Routines in a programming language perform three functions:

1. Provide modularity.
2. Divide the program into abstraction levels.
3. Reduce code duplication.

6.2 Constraints of procedural programming

The constraints are 3 additional principles of procedural programming.

6.3 Procedural and imperative programming

Imperative programming is a paradigm whereby a program is a sequence of commands that change its state. Structured programming and procedural programming are kinds of imperative programming. However, in some books (for example, [19, 20]), procedural programming is a synonym for imperative programming, which differs from the paradigm of procedural programming used in this paper and, for example, in [3, 21].

Procedural programming had the only way to reduce code duplication. It is the division of the program's text into routines, each of which solves a specific function. New ways have appeared in the paradigm of object-oriented programming.

7. Object-Oriented Programming

7.1 The Fundamentals of Object-Oriented Programming

Object-oriented programming appeared at the same time as structured. For example, E. Dijkstra's ideas about structured programming and the description of the Simula object-oriented programming language were in various chapters of one book [4]. However, object-oriented programming was not claimed, as the possibilities of structured programming and procedural programming were not exhausted. Object-oriented programming slowly developed and it became the dominant programming paradigm in the 1990s. Indeed new ways to reduce code duplication allowed object-oriented programming to crowd out other paradigms.

Object-oriented programming is based on three basic principles: inheritance, encapsulation, and polymorphism. Each of these principles reduces code duplication.

7.2 Inheritance

«Inheritance is also fundamental in gathering commonalities and removing duplications» [9].

«In principle, inheritance can be implemented by code duplication. For every objector class of objects defined by inheritance, there is a corresponding definition that does not use inheritance; it is obtained by expansion of the definition so that inherited code is duplicated. The importance of inheritance is that it saves the effort of duplicating (or reading duplicated) code and that, when one class is implemented by inheriting from another, changes to one affect the other. This has a significant impact on code maintenance and modification» [22].

7.3 Polymorphism

Polymorphism (with dynamic binding) allows reducing code duplication in two aspects:

1. When calling a polymorphic method, it is not necessary to check its belonging to a class (Listing 1). The dynamic binding mechanism does this automatically.

Listing 1. Polymorphic method call with class belonging check

```
if (object instanceof ClassA) {
        ((ClassA) object).someMethod();
}
```

2. Polymorphism is used to reduce code duplication in some design patterns, for example, to get rid of duplication of case values in the switch statements [7, 8].

7.4 Encapsulation

Encapsulation was born in the structured programming times as information hiding [23]. So we consider encapsulation, which specific to object-oriented programming: restrict an access to instance variables in a class only by special methods.

Encapsulation reduces code duplication due to the fact that the rules for accessing the field are implemented only in special methods, and not with each reference to the field.

7.5 Constraints of Object-Oriented Programming

Constraints of object-oriented programming:

1. Describe the solution of the problem by classes and their instances.

2. Classes and their interrelations must comply with the design principles: the Single-Responsibility principle, the Open-Closed principle, the Liskov Substitution principle, the Interface-Segregation principle, the Dependency-Inversion Principle [24].

3. Adequately use inheritance, encapsulation, and polymorphism to simplify program maintenance.

4. When designing a software system and a class system, use design patterns [25].

Note the refactoring of the program's class system also reduces code duplication [7].

8. Summary

We considered ways to reduce code duplication and constraints of the programming paradigms (see the Table). Each paradigm came from the previous one, so it inherited all existing ways to reduce duplication and constraints.

9. Conclusions

We have made the following conclusions.

I. We have analyzed the imperative programming paradigms: structured, procedural, and object-oriented for compliance with the criterion for choosing the structure of the program. For each paradigm, ways to reduce code duplication were discovered. We have concluded the evolution of programming paradigms was influenced by the search for new ways to solve the problem of code duplication.

II. Despite the appearance in programming languages of new ways to reduce code duplication, the problem remains actual. The reason for the appearance of duplication is not the shortcomings of a programming language or software that eliminates it, but a low qualification of programmers. Therefore, the main way to reduce code duplication is to focus attention on this problem during programming teaching, its negative consequences and ways to reduce it, which exist in modern programming languages.

III. If the possibilities of reducing duplication are exhausted, then it will be necessary to choose another way to increase the degree of compliance with the criterion and the evolution of programming languages will follow a new way.

IV. We have listed the constraints introduced by programming paradigms. The introduction of constraints allowed the development of more complex software systems. We can't develop more complex system by more simple programming tools.

V. The evolution of the programming paradigms considered in the paper influenced other paradigms (for example, functional [21]).

Avoiding code duplication is an obvious way to reduce the expenses of developing a program: write a program faster. However, with the paradigm of object-oriented programming, new ways to reduce expenses have arisen: encapsulation of program details, design patterns, etc.

«Every time you see duplication in the code, it represents a missed opportunity for abstraction. That duplication could probably become a subroutine or perhaps another class outright. By folding the duplication into such an abstraction, you increase the vocabulary of the language of your design. Other programmers can use the abstract facilities you create.

Coding becomes faster and less error prone because you have raised the abstraction level» [8].

Table. Ways to reduce code duplication and constraints of structured, procedural, and object-oriented programming.

Programming paradigm	Ways to reduce code duplication	Constraints of the paradigm
Structured programming	Routines, but they were not used	1. Use only sequence, selection and repetition control constructs in programs. 2. Develop a program using step-wise composition.
Procedural programming	Routines	1. Functions of each level of abstraction are described by routines. 2. Each routine performs one function at its own abstraction level. 3. The routine is single-entry, single-exit control construct, receives the input data through parameters and returns the result through parameters or through the return statement.
Object-oriented programming	Inheritance, encapsulation, and polymorphism	1. Describe the solution of the problem by classes and their instances. 2. Classes and their interrelations must comply with the design principles: the Single-Responsibility principle, the Open-Closed principle, the Liskov Substitution principle, the Interface-Segregation principle, the Dependency-Inversion Principle. 3. Adequately use inheritance, encapsulation and polymorphism to simplify program maintenance. 4. When designing a software system and a class system, use design patterns.

References

[1] Finkel R A 1996 *Advanced Programming Language Design* (Menlo Park: Addison-Wesley)

[2] Booch G, Maksimchuk R A, Engle M W, Young B J, Conallen J and Houston K A 2007 *Object-Oriented Analysis and Design with Applications* (Redwood: Addison-Wesley)

[3] Istatkova G 2006 *Proc. of Int. Conf. on Computer Systems and Technologies – CompSysTech*

[4] Dahl O-J, Dijkstra E W and Hoare C A R 1972 *Structured programming* (London: Academic Press)

[5] Fontana F A, Zanoni M and Zanoni F 2015 *Agile Processes in Software Engineering and Extreme Programming eds C Lassenius et al* (Berlin: Springer) pp 3–14

[6] Koschke R 2008 *Software Evolution eds T Mens and S Demeyer* (Berlin: Springer) pp 15–36

[7] Fowler M, Beck K, Brant J, Opdyke W and Roberts D 1999 Refactoring *Improving the Design of Existing Code* (Menlo Park: Addison-Wesley)

[8] Martin R 2009 *Clean Code. A Handbook of Agile Software Craftsmanship* (New York: Prentice Hall)

[9] Meyer B 1997 *Object-Oriented Software Construction* (Englewood Cliffs: Prentice Hall)

[10] Mills H D 1971 *Debugging Techniques in Large Systems ed R Rustin* (Englewood Cliffs: Prentice Hall) pp 41–55

[11] Yourdon E 1975 *Techniques of Program Structure and Design* (Upper Saddle River: Prentice Hall)

[12] Dijkstra E W 1969 *Software Engineering Techniques eds J N Buxton and B Randell* (Brussels: NATO Scientific Affairs Division) pp 84–8

[13] Mills H D 1988 *Software productivity* (New York: Dorset House)

[14] Baker F T and Mills H D 1973 *Datamation* **19** 58–61

[15] Hugo I ST J 1977 *Proc. of AFIPS'77* pp 741–52

[16] Stein J 1988 *Dr. Dobb's Journal of Software Tools for the Professional Programmer* **137** 18

[17] Dijkstra E W 1968 *Communications of the ACM* 11 pp 341–6

[18] Liskov B H 1972 *Proc. of Fall Joint Computer Conference* pp 191–9

[19] Brookshear J G 2012 *Computer Science. An Overview* (Reading: Addison-Wesley)

[20] O'Regan G 2012 *A Brief History of Computing* (London: Springer)

[21] Hyvönen E and Seppänen J 1987 *2 Lisp-Maailмa. Ohjelmointimenetelmat ja -jarjestelmat* (Helsinki: Kirjayhtymä Oy)

[22] Mitchell J C 2002 *Concepts in Programming Languages* (New York: Cambridge University Press)

[23] Parnas D L 1972 *Communications of the ACM* **15** 1053–8

[24] Martin R 2003 *Agile Software Development: Principles, Patterns, and Practices* (Upper Saddle River: Prentice Hall)

[25] Gamma E, Helm R, Johnson R and Vlissides J 1995 *Design Patterns Elements of Reusable Object-Oriented Software* (Reading: Addison-Wesley)

Investigation of the Influence of the Cutout Dimensions on the Stress-strain State of Three-layer Shells with Load-bearing Layers of Composite Materials

V N Bakulin[1,2]

[1] Moscow Aviation Institute (National Research University), Moscow, Russia
[2] Institute of Applied Mechanics, Russian Academy of Sciences, Moscow, Russia

E-mail: vbak@yandex.ru

Received xxxxxx
Accepted for publication xxxxxx
Published xxxxxx

Abstract

The description of the model for investigation of the influence of rectangular dimensions in terms of cutouts on the stress-strain state (SSS) in the layers of three-layer conical shells is given. The finite element model is based on the layer-by-layer analysis approach and allows to perform calculations of SSS in shell layers in a wide range of changes in the physical and mechanical characteristics of the layers in the presence of cut-outs. The equations of the classical theory of shells at reception of effective approximations (leading to high speed of convergence of received results) and construction of finite-element model of moment bearing layers are used. The approach with application of the elasticity theory relations for a three-dimensional body in curvilinear coordinates and obtained approximations for construction of a finite element model of the filler layer is described. Influence of the cut-out mortar angle on the parameters of the three-layer shell with load-bearing layers of composite materials is investigated.

Keywords: three-layer conical shells, composite materials, rectangular cut-outs, finite element model, stress-strain state

1. Introduction

When designing aviation and aerospace equipment, three-layer shells [1], including composite materials, are spread. This is due to the high weight efficiency and rigidity, high thermal, acoustic and vibration resistance of three-layer structures. An example of the application of three-layer and especially cellular structures can serve as a supersonic bomber firm "Convair" B-58. The surface of the glider of this aircraft (80–90% of the total streamlined area) is made of structures with filler. These structures are used in the power elements of the wing, fuselage.

The Dynamic-Grumman F-111B aircraft has about 300 cellular panels (basically all nodes except the nose fairing, the small area behind the cockpit, the middle areas of the wing and the keel).

The construction with the filler of C-141 and C-5A aircrafts by Lockheed is of certain interest. The total area occupied by cellular panels on C-141 aircraft is 557 м^2.

Experience of manufacturing and operation of panels with the filler of C-M1 aircraft was considered by the company in the construction of the American strategic military transport aircraft C-5L "Galaxy". The total area of cellular structures on this aircraft is increased by 4.5 times and is 2320 $м^2$ with an average panel thickness of 25.4 mm.

Filler panels on C-5L aircraft are used in the following designs: nose and tail parts of the wing and nose parts of the fuselage, tail part of the flaps, engine pylons, cargo hatches, elements of the tail part of the fuselage, fairings of the main landing gear, etc.

Structures with filler are also used in space equipment for power units of apparatuses and heat-insulating screens.

Aeroika's multi-layer design, used for atmospheric re-entry vehicles, deserves consideration. The multi-layer filler ensures a high thermal insulation coefficient with a minimum weight of the structure.

Aircraft and aerospace engineering structures need to have different purpose cut-outs, hatches that weaken the load-bearing capacity. This explains the continuous improvement of methods for calculating the stress-strain state (SSS) of structural elements, including three-layer shells, with cut-outs.

2. Description of the research

Conical shells, including three-layer structures, are widely used in the design of aviation and aerospace equipment. This work is devoted to the influence of the rectangular solution angle on the stress-strain state in the layers of three-layer conical shells. It should be noted that the problem of analyzing the stress-strain state of three-layer shells, considering the heterogeneity of the structure and the presence of cut-outs, rectangular in terms of shape, is insufficiently studied.

There is a lot of work to build theories of layered and three-layer shells. Among them are the works of S.A. Ambartsumian, V.V. Bolotin, V.V. Vasiliev, N.K. Galimov, E.I. Grigolyuk, Y.M. Grigorenko, M.A. Ilgamov, H.M. Mustari, V.N. Paymushin, S.N. Sukhinin, E. Reissner, L. Hebeep and many others.

In order to calculate the VAT of three-layer shells with cut-outs under different conditions of fixing and loading, variable properties of layers, it is necessary to develop models based on numerical methods and, first, the finite element method (FEM). Monographs are devoted to the construction of finite element models (FEM) of layered and three-layer shells [2–6].

In most works, finite element models are based on hypotheses for a package of layers. More general and accurate, but less common, is the approach where hypotheses are applied to each layer. It makes actual development of the refined models based on the layer-by-layer analysis, allowing to solve the marked problems and to carry out calculations of the intensely deformed condition of three-layer shells with cut-outs. At the layer-by-layer analysis the order of equation systems depends on the number of layers, which considerably increases the dimensionality of matrices and complicates the solution of problems. Therefore, when building refined models based on layer-by-layer analysis, it is necessary to pay special attention to the basic functions of approximations, on the basis of which the finite elements are built [7]. Usually, elements with basic functions based on polynomial approximations of non-high orders of magnitude are used to calculate heterogeneous shells using the finite element method. At application of finite element models with such basic functions at the layer-by-layer analysis of the stress-strain state of three-layer shells with cut-outs it is necessary to use a considerable quantity of finite elements that will lead to increase in dimensionality of the matrixes participating in the calculation process.

In works [8, 9] at construction of finite elements for calculation of homogeneous shells it has been shown that speed of convergence of numerical results considerably increases, and consequently, the order of systems of the equations decreases, if in approximation of displacements to enter displacements as a solid [10, 11]. It has allowed to receive more exact results at smaller number of finite elements in comparison with use of finite elements without considering movements as a solid body. Such an increase in the rate of convergence of numerical results due to the construction of effective approximation functions and their application in the finite elements is especially relevant for the layer-by-layer analysis [12] of the stress-strain state of three-layer shells with cut-outs, as it will lead to a decrease in the dimension of matrices involved in the calculation process. This will have a positive impact on the accuracy of calculations and reduce the computing requirements.

3. Model for investigation of the influence of rectangular cuts on the stress-strain state in layers of three-layer conical shells

The refined finite-element model for layer-by-layer analysis is considered, which allows to carry out calculations of the stress-strain state in layers of three-layer conical shells in a wide range of changes in the geometric and physical-mechanical characteristics of the layers in the presence of cut-outs. The methodology of finite-element model construction is based on the development of the layer-by-layer analysis approach for the creation of block models of three-layer, in general, irregular shells; the use of equations and relations of the classical theory of shells in obtaining effective finite-element approximations (FEA), resulting in high convergence rate of numerical procedures and obtained results, and construction of finite element model of natural curvature for momentary bearing layers; the proposed approach with application of elasticity theory for three-dimensional body in curvilinear coordinates and effective FEA in the construction of FEM of natural curvature for the filler layer. Thus, the simulation of the stress-strain state in the bearing layers will be carried out with the help of two-dimensional cladding finite elements (FE) and with the help of three-dimensional cladding FE modeling of SSS in the filler layer.

Approximating functions of displacement of the median surface of the finite elements of the bearing layers are composed of displacements as a solid body and displacements caused by deformation of FE. Displacements as a solid body are determined by the integration of Cauchy relations [13] at zero deformation values and are recorded similarly [7, 10, 14] through six undefined coefficients, which are constant integration. The efficiency of approximations leading to a high rate of convergence of numerical procedures and obtained results, and the constructed finite element model of natural curvature for the moment-bearing layers of three-layer conical shells are shown in [12].

A block of two-dimensional and three-dimensional shell finite elements is assembled from the thickness of the three-layer shell (block three-layer conical element), obtained by the cross section of two planes passing through the shell axis and two conical surfaces formed by rotation around the shell axis to the median surface at the node points of FE.

Application of two-dimensional and three-dimensional envelope finite elements for modeling accordingly momentary bearing layers and a layer of a filler will lead to rupture of the generalized moving on contact surfaces of these finite elements if they are constructed with use of various approximating functions of moving. In order to avoid errors caused by such a gap in generalized displacements, the same number of nodes is selected for the finite elements of the filler layer on conical surfaces as for the finite elements of the bearing layers, and the same generalized displacements and approximating functions as for the finite elements of the bearing layers are used as nodal unknown and approximating functions. As the nodal generalized displacements for the finite elements of the bearing layers and, consequently, the finite elements of the filler layer, there are three linear displacements and two normal rotation angles to the median surface around the meridional and circumferential coordinate axes in each of the four knots of the bearing layers' FE and eight knots of the finite elements of the filler layer.

The displacements caused by deformation of the finite elements of the bearing layers are approximated by an incomplete bicubic polynomial deflection polynomial and bilinear polynomial deflection polynomials for meridional and circular displacements. Having obtained the approximating functions of the displacements of the finite elements of the bearing layers (recorded with twenty undefined coefficients) and substituting them for the Cauchy relations linking strains with displacements, we express the dependences for generalized strains through these twenty undefined coefficients.

The Cauchy relations that bind deformations with displacements for a three-dimensional body in curvilinear coordinates [15] are applicable for the final element of the filler layer, which is a thick-walled conical shell

$$\varepsilon_{11} = \partial_x u,\ \varepsilon_{22} = r^{-1}(\partial_\beta v + u\sin\gamma + w\cos\gamma),\ \varepsilon_{33} = \partial_z w,$$

$$\gamma_{12} = r^{-1}\partial_\beta u + \partial_x v - r^{-1}v\sin\gamma,\ \gamma_{23} = r^{-1}\partial_\beta w + \partial_z v - r^{-1}v\cos\gamma,$$

$$\gamma_{31} = \partial_x w + \partial_z u,$$

where $r = R_2\cos\gamma, R_2 = (R_1 + R_3)/2$, R_1 and R_3 radius of the inner and outer conical surfaces of the final element of the filler layer, respectively.

To obtain approximating displacement functions of the finite elements of the bearing layers on the surfaces of contact with the filler layer, we use the transition matrix from the middle surface of the bearing layers to the interfaces with the filler layer in a similar way [16].

Having determined the approximating functions of the movement of the finite elements of the bearing layers, and therefore approximating functions of the movement on the conical surfaces of the finite element of the filler layer, and applying the linear law of displacement change in the thickness of the final element of the filler layer from (1), we obtain the dependences for generalized strains, recorded through forty undefined coefficients.

Knowing the relations for the physical law for the bearing layers and the filler layer, with the help of the obtained dependencies of generalized deformations we write down the expressions for stresses in the finite elements of the layers of the block three-layer conical element. Further, using procedures such as [17], we define the matrixes of finite element stiffness of the layers and proceed to the construction of a finite element model of a three-layer conical shell.

4. Investigation of the influence of rectangular cutouts on the stress-strain state of the three-layer shell

As noted above, the problem of the analysis of the stress-strain state of three-layer shells, considering the heterogeneity of the structure and the presence of rectangular cut-outs in terms of shape, has not been sufficiently investigated. Not many works in this direction are known, for example, [12, 18, 19]. Obtaining an analytical solution to study the stress-strain state of three-layer shells, taking into account the heterogeneity of the structure and the presence of cut-outs, rectangular in terms of shape, is associated with great difficulty. This results in the application of numerical methods and, first of all, the finite element method.

Let's consider the problem about the influence of rectangular cutout sizes on the stress-strain state of the three-layer conical shell. A three-layer conical compartment with two diametrically opposed rectangular cut-outs, loaded with uniformly distributed internal pressure, was chosen as the object of study. The bearing layers are made of fiberglass and the filler is made of polyurethane foam. The boundary conditions at the ends of the casing are considered to correspond to the case of a rigid seal (axial movement is permitted at the end of the smaller diameter).

The geometric parameters of the shell are as follows:

$$R = 1.5 \text{ m}, \ L = 1.5 \text{ m}, \ h_1 = 0.43 \text{ cm}, \ h_2 = 0.21 \text{ cm},$$

$$H = 5.63 \text{ cm}, \ \gamma = 20.556^\circ,$$

where R — inner radius of the larger base; L — bay height; h_1, h_2 — the thickness of the inner and outer load-bearing layers accordingly; H — three-layer package thickness, γ — taper angle.

The cutouts are located at an equal distance from the ends of the casing. The length of the cutouts is 1/3 of the length of the cutouts forming. Cutout corner angle α_B was changing within the limits of $30^\circ \le \alpha_B \le 60^\circ$.

The physical and mechanical characteristics of the three-layer shell are as follows:

- for the inner bearing layer: $E_1 = 2.5 \cdot 10^5$ kg/cm², $E_2 = 2.2 \cdot 10^5$ kg/cm², $G_{12} = 0.35 \cdot 10^5$ kg/cm², $\mu_2 = 0.1$;
- for the outer load-bearing layer: $E_1 = 2.1 \cdot 10^5$ kg/cm², $E_2 = 1.9 \cdot 10^5$ kg/cm², $G_{12} = 0.36 \cdot 10^5$ kg/cm², $\mu_2 = 0.1$;
- for the filler: $E_{11} = E_{22} = E_{33} = 240$ kg/cm², $G_{13} = G_{23} = G_{12} = 100$ kg/cm².

Due to symmetry, the calculation considers the 1/4 symmetrical part of the shell, which was divided into 24 elements in the axial direction and 30 elements in the circumferential direction.

Fig.1, 2 shows graphs of changes in the main parameters of the stress state of membrane tensions σ_{N1}, σ_{N2} in the bearing layers along the meridional (Fig. 1) and circumferential (Fig. 2) coordinates for two values of the cutout angle: $\alpha_B = 30^\circ$ and $\alpha_B = 60^\circ$. Figures 1, 2 in the figures show graphs of voltage distribution in the inner and outer load-bearing layers, respectively, for the case $\alpha_B = 30^\circ$, with the numbers 3, 4 for the case $\alpha_B = 60^\circ$. Fig. 1 shows diagrams in the section of the plane passing through the shell axis and the meridional line located at a distance of half the size of the finite elements in the circumferential direction from the rectilinear edge of the cut (let's mark this line A — A), where N_{AX} — FE numbers on the line A — A, counted from the base of the large-diameter cone.

Fig. 2 shows diagrams of the cross-section of the plane perpendicular to the axis of the shell and passing through a line parallel to the curvilinear edge of the large-diameter cutout and located at a distance of half the size of the finite elements from this edge of the cut (let's mark this line B — B), where $N_{B\beta}$ — FE numbers on the line B — B, the countdown is taken from the symmetry plane passing through the middle of the curved edges of the cut.

As can be seen from the above graphical dependences, the stress-strain state of the shell weakened by the cutouts is characterized by a pronounced edge effect in the vicinity of the corner points of the cutouts, which quickly fades away as the distance from them. The maximum values of membrane and torque stresses in this problem are observed in the internal bearing layer. In the outer load-bearing layer, the maximum stress values are considerably lower than in the inner load-bearing layer. The largest absolute values in the bearing layers are diaphragm stresses σ_{N1}, σ_{N2}. The maximum values of shear voltages are 2–2.5 times less than the maximum stresses σ_{N1}, σ_{N2}.

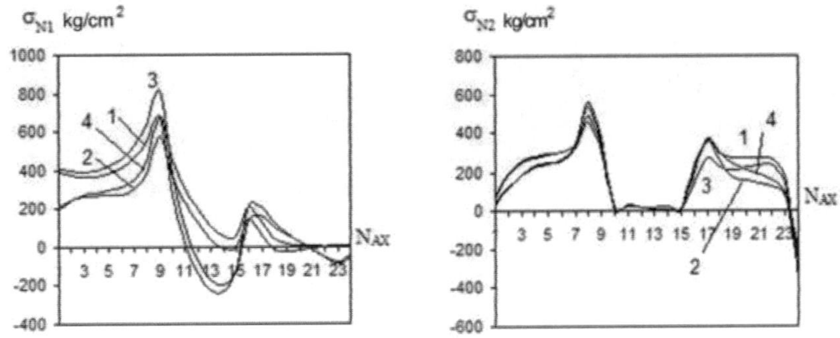

Figure 1. Diagrams of diaphragm stress changes σ_{N1}, σ_{N2} in the bearing layers along the meridional coordinate in the section by the plane separated from the rectilinear edge of the rectangular cutout at a distance of half the size of the FE, for two cutout angles: 1, 2 in the inner and outer support layers, respectively, for the case $\alpha_B = 30^\circ$, 3, 4 — in case of $\alpha_B = 60^\circ$.

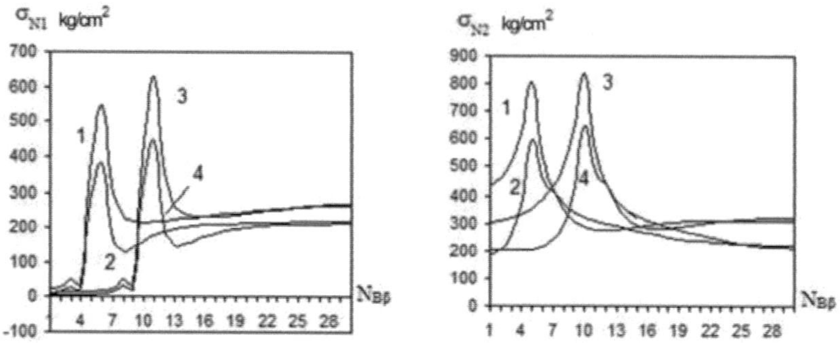

Figure 2. Diagrams of diaphragm stress changes σ_{N1}, σ_{N2} in the bearing layers at the circumferential coordinate in the section of the plane separated from the curved edge with a large diameter of a rectangular cut at a distance of half the size of the FE, for two cutout angles: 1, 2 in the inner and outer support layers, respectively, for the case $\alpha_B = 30°$, 3, 4 — in case of $\alpha_B = 60°$.

It should be noted that the maximum voltage values in the filler are σ_{11}, σ_{22}, which in most cases are neglected, are comparable to the maximum values of voltages σ_{33}, τ_{31}, τ_{23}, usually taken into account in the calculations. And only for stresses τ_{12} the maximum values are several times lower than for the other components of the core stress state.

Fig. 3 shows graphical dependences of the stress state parameters in the inner bearing layer near the corner point of the cut on the size of the cut. Figures 1 to 6 indicate the stress dependencies σ_{Tx}, $\sigma_{T\varphi}$, σ_S, σ_{Mx}, $\sigma_{M\varphi}$, σ_H, respectively. The presented parametric dependences show that the change in the cutout size within the specified limits has little effect on the maximum sheath stress values. So when you increase the angle of the cutout by a factor of 2 (from 30° to 60°) maximum values of diaphragm stresses σ_{N1}, σ_{N2} in the inner bearing layer are increased by 5–15%.

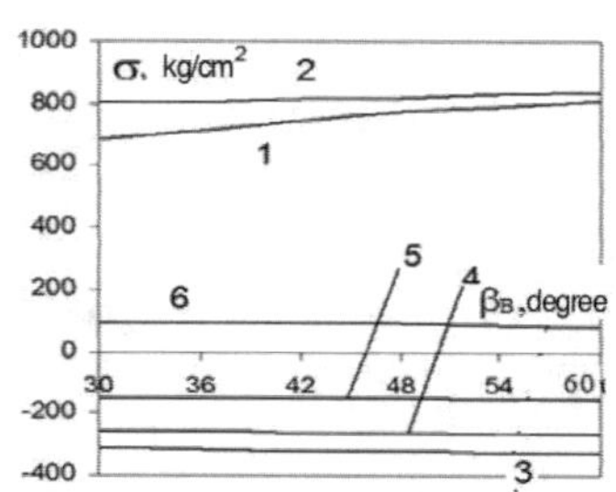

Figure 3. Dependence of the stress state parameters in the inner bearing layer near the corner point of the cut on the cutout dimensions. The numbers 1–6 correspond to the voltages $\sigma_{N1}, \sigma_{N2}, \sigma_S, \sigma_{M1}, \sigma_{M2}, \sigma_H$.

Thus, from the given results it is possible to draw a conclusion that change of an angle of a solution of a cut out in the specified limits practically does not render appreciable influence on parameters of the SSS of a three-layer conic cover.

Acknowledgments

The work was done within the framework of the state assignment, the subject of the state registration number AAAA-A19-119012290177-0.

References

[1] Panin V F and Gladkov Y A 1991 *Structures with Filler: Handbook* (Moscow: Mechanical Engineering)
[2] Alfutov N A, Zinoviev P A and Popov B G 1984 *Calculation of multilayer plates and shells from composite materials* (Moscow: Mechanical Engineering)
[3] Rasskazov A O, Sokolovskaya I and Shulga N A 1986 *Theory and calculation of layered orthotropic plates and shells* (Kiev: Vischa School)
[4] Piskunov V G, Verizhenko V E, Prysyazhnyuk V K et al. 1987 *Calculation of heterogeneous flat shells and plates by finite element method* (Kiev: Vischa school)
[5] Rickards R B 1988 *Finite element method in shell and plate theory* (Riga: Zinatne)
[6] Kaledin V O, Aulchenko S M, Mitkevich A B, Reshetnikova E V, Sedova E A and Shpakova Yu V 2014 *Modeling of statics and dynamics of shell structures from composite materials* (Moscow: Fizmatlit)
[7] Bakulin V N 2007 *Doklady Physics* **6** 330–4
[8] Cantin G and Glagh R W 1968 *AIAA Journal* **6** 1057–62
[9] Ashwell D G and Sabir A B 1972 *Int. J. of Mechanical S.* **3** 171–83
[10] Demidov V I and Bakulin V N 1978 *Izvestia Machinostr* **5** 5–10
[11] Zheleznov L P and Kabanov V V 1990 *Izvestia RAS. MTT* **1** 131–6
[12] Bakulin V N 2018 *Doklady Physics* **1** 23–7
[13] Novozhilov B B 1951 *Theory of Thin Shells* (Leningrad: Sudpromgiz)
[14] Bakulin V N 2019 *Doklady Physics* **1** 9–13

[15] Vasilyev B B 1988 *Mechanics of composite materials constructions* (Moscow: Mechanical Engineering)
[16] Krivtsov V S, Rassokha A A and Bakulin V N 1983 *Izvestia VUZov. Aviation engineering* **4** 14–8
[17] Bakulin V N 2006 *Matem. Mod.* 1 3–9
[18] Aksentian K B and Krasnobaev I A 1973 Basic bending equations and method of calculation of circular three-layer cylindrical shell with a large rectangular cutout. *The theory of shells and plates* (Moscow: Science) pp 601–5
[19] Aksentian K B and Krasnobaev I A 1973 *Izvestia VUZov Construction and architecture* **2** 45–51

The Modified Evolutionary Algorithm for the Algorithms' Chain Formation that Improve Gradient Images

V V Belov[1] and A K Lopatin[1,2]

[1] Department of Computational and Applied Mathematics Federal State Budgetary Educational Institution of Higher Education "Ryazan State Radio Engineering University named after V. F. Utkin", Ryazan, Russia
[2] State Educational Institution of Higher Education of Moscow Region "State University of Humanities and Social Studies", Kolomna, Russia

E-mail: ak_lopatin@mail.ru

Received xxxxxx
Accepted for publication xxxxxx
Published xxxxxx

Abstract

The problem of improvement of images, in implementation process of automatic control of the geometrical sizes of the products which are on the conveyer belt is considered. The purpose of work is development of the method of automatic structural and parametrical identification of the combination of algorithms for the solution of the problem of preprocessing of images. Serve as input parameters of the system: the image of the controlled product in the RGB format and the image standard in the bitmap. The algorithm of correction of defects of the image and the subsequent selection of edges is offered adapting to shooting conditions (camera type, its permission and surface illumination). On the output the vector of operators and their parameters forms.

Keywords: genetic algorithms, image processing, compositions of algorithms, optical checking, set of algorithms, filtering of images, selection of edges

1. Introduction

Many algorithms of a solution of the tasks anyway connected with image processing [1] is so far created, but all of them have one general beginning: all sequence of these algorithms steps — from initial prior to a final procedure — is a product of natural (human) intelligence. The complexity of solvable tasks, variety and variability of conditions of observation of problem objects more and more persistently stimulate requirement in accelerated, and, therefore, automatic synthesis of image processing algorithms, including determination of structure of necessary procedures and their parameters.

In the real work results of researches which incentive to carrying out was requirement of one of the industrial enterprises occupied with production of details of aerospace assignment for implementation of automatic measurement of overall dimensions of the products which are on the conveyer belt on their pictures are described. The task of creation of a system capable to provide automatic adaptation of process of measurement to fluctuations of conditions of observation of objects in the conditions of real production was set. Within the specified task, in particular, the stated technique of forming of the gradient images applied further to assessment of quality of products — correctness of their geometrical sizes and lack of unacceptable damages of a surface was created.

Development of a technique was preceded by the wording of the design requirement, typical for program systems: the synthesizable chain of operators should be optimum by two criteria: it should combine the high accuracy of recovery of a circuit of a product with a labor input minimum.

2. Analysis of experience of automatic synthesis of algorithms of image processing

Issues of automation synthesis of image processing algorithms are presented in publications poorly. The approaches to automation images quality improvement applied now are stated in [2–7]. It is possible to select two basic approaches:

1) application of genetic algorithms [2–4];

2) application of artificial neural networks (ANN) [6].

2.1 Experience of synthesis of chains of standard algorithms in practice of improvement of images

In [2–4] genetic algorithm it is used for automation of process of synthesis of an algorithm of image processing, received as a result of ultrasonic sounding. The required algorithm is synthesized by the genetic algorithm in the form of the algorithms composition implementing some basic procedures (operators). The purpose of synthesis is receiving the algorithms' chain providing image enhancement before achievement of the maximum approach to the standard image of an object submitted on the image.

Degree of proximity of images is evaluated by value of the fitness function calculated in the course of pixel-by-pixel comparison. It is claimed that fitness function can have different definitions, but specifically it is defined once as follows:

$$Fitness = \min\left(\frac{|Q \cap Q'|}{|Q \cup Q'|}, \frac{|P \cap P'|}{|P \cup P'|}\right), \qquad (1)$$

where Q, Q' — sets of bright pixels, P, P' — sets of black pixels of the reference and compared fragments of images respectively.

The genetic algorithm is applied as a method of confirmation of optimality of the sequence of the image processing offered by the author based on intuitive reasons. The analysis of several combinations of basic algorithms is carried out and depending on function value of fitness (fitness function) on different sets of input data the conclusion about advantage of the processing technique offered by the author is drawn. In [3] important, though rather obvious output [3] pages 78 contains: "For some new image, without setting a template, it is impossible to select arbitrarily the sequence, from earlier constructed GA and to expect required result of processing" (here GA — "genetic algorithm").

In relation to a problem of products automatic measurement this output means that the composition of algorithms which is rather well improving the image of one product does not guarantee so good image enhancement of other product. For each type of a product (the type of a product is called a template, a standard or a sample) it is required own structural and, what is much more often, — parametrical identification of algorithms of image processing. Reveal what composition of algorithms needs to be used at the moment, it is possible in two ways: 1) a solution of a recognition problem of a product type on the image; 2) exogenous indication of necessary information, inclusion of an observed product type in structure of an algorithm's basic data. The first way is applicable anyway, and the second — only when types of observed products change seldom, or different products are observed on different conveyer belts.

Among features of the technique offered in [3] we will note a potentiality of repeated application of the same algorithm to the same image and any arrangement (generally — any arrangements) algorithms in the created chain. In [3] it is claimed that this feature allows to receive necessary result — rather high similarity of a standard and result of image processing. However, as practice shows, the specified statement is fair not always.

In the context of the considered task it is possible to specify the following of a technique shortcomings [3–4].

1. The disorder and potential recurrence of algorithms in the created chains leads to the fact that the genetic algorithm, carrying out search of possible chains, creates and evaluates the options containing repeated repetitions opposing on the nature of impact on the image of operators. For example, chains can form and repeat: "smoothing (filtering, integration) — selection of borders (calculation of a gradient, differentiation) — smoothing — selection of borders". In tasks in which the visual similarity of results of image processing to their reference options not essentially and matters only the accuracy of selection of circuits repetition of filtering and selection of borders is deprived of sense and leads to unjustified delay in a procedure of search of an optimum chain of algorithms. Tightening can be so essential that there is a need for intervention of experts for the purpose of truncation of the created chains.

2. Orientation to maximizing similarity of result of image processing to a standard can lead to removal of target information: the violations of a form, a crack and a cavity which are an identification subject during automatic quality control of products can be deleted.

3. Preliminary determinancy of algorithms' parameters values. As an argument in favor of predetermined parameters serve experience and expert opinions. Partly it is proved also by practical experiments [8, 10], but at the same time there are reasons for which changes of parameters of algorithms, for example, of the radius of a window of the filter or a threshold of the operator of selection of borders can be required:

• change of the fixing equipment model;

• change of shooting conditions;

• microdefects of the specific camera.

Specified means that generally parametrical identification of algorithms is the necessary procedure.

4. Rather significant participation of the expert in the course of forming of an optimum chain of algorithms. This feature is unacceptable if process of forming of a new chain of algorithms becomes a part of technological process of day-to-day running of the software.

2.2 Experience of improvement of quality of images based on function of recalculation of brightness of pixels

The idea of a method [2] which is intensively used in numerous works of authors [3–5] consists in the repeating recalculation of the current values of array elements of the image pixels' brightness in new it is expected the best values. For recalculation the function called by an improvement kernel with some set of parameters is used. In [2] number of parameters of a kernel of improvement it is equal to four.

Values of parameters are selected by the genetic algorithm means. As a matter of fact, recalculation of a matrix of pixels brightness is implemented as a component of an optimization problem solution in which target function is image quality and required variables — improvement kernel parameters. Repetition of pixels brightness recalculations is carried out until execution of genetic algorithm's stop criterion — extremely seldom end is carried out because of a global, or suboptimal solution achievement, much more often the stop happens because of number of the generations exhaustion which are released on evolution or exhaustion of time which is released on evolution.

Target function represents convolution of three private criteria. In [2] three private criteria are offered: 1) number of boundary pixels; 2) intensity of edge pixels; 3) measure of entropy of the image. For calculation of the first two criteria the discrete differential Sobel operator calculating approximate value of a gradient of image brightness is used. Authors [5–7] added one more to the specified three private criteria: the adaptation layer to sight of the person on brightness.

Thus, in the algorithms developing the idea of a method [2], genetic algorithms are used as an alternative to the standard procedures of images' quality improvement having the accurate mathematical description. Though the total failure does not come from standard procedures: 1) for calculation of two private criteria the standard operator of calculation of a gradient of image brightness is used (of course, it is possible to use not only the Sobel operator [8], but also any other); 2) some modifications [7] assume two-stage solutions — at the first stage the standard image enhancement procedure due to alignment of the histogram is implemented, and on the second the brightness recalculation algorithm in the context of a optimization problem's solution by means of the genetic algorithm is started [3].

2.3 The offered technique of synthesis of chains of the algorithms improving quality of gradient images

In the course of synthesis of experience of process automation of structural and parametrical identification of the image processing algorithms following working hypotheses were formulated.

1. There is a certain "reasonable" sequence of steps (functions, private tasks) in the course of a solution of a problem of improvement of gradient images of the products presented in the processed pictures. Each step is an application of this or that operator (algorithm) of image processing of specific functionality.

This sequence of steps is as follows:

1) alignment of the histogram of the image;

2) filtering (smoothing) of the image;

3) selection of borders by calculation of a gradient, i. e. obtaining the gradient (planimetric) image;

4) specification of borders — the process of minimization of thickness of circuits having numerous alternative ways of implementation, the simplest of which represents set of the following procedures:

4.1) closing — the consecutive application of dilatation (expansion) and an erosion (corrosion) leading to removal in circuits of gaps up to two pixels long;

4.2) removal of single pixels — pixels equal 1, all 8 neighbors of which are equal 0, are replaced with 0.

Use of a specific way of specification of limits of basic value has no.

2. Repeated application of any of the specified steps, except filtering, is inexpedient, that is does not promote increase in accuracy of recovery of a circuit of a product. Moreover, repeated application of operators of selection of borders is destructive, – in particular, leads to emergence of repetitions of circuits.

3. Natural trend: than the algorithm is more difficult, especially it provides high quality of image processing (in particular, operators of selection of borders with bigger labor input provide more exact recovery of circuits of objects according to their photos) when using chains of algorithms is shown poorly, — is practically absent.

4. At a solution of a specific applied objective steps of image processing can be divided into obligatory and optional. In the course of improvement of quality of gradient images, the step of selection of borders is obligatory, all other steps are optional, i.e. cannot be carried out (at high quality of photos).

The specified hypotheses were never disproved in the course of the numerous experiments with images of the products which are subject to control during their movement on the conveyer belt. Based on these hypotheses the technique of synthesis of chains of the algorithms improving quality of

IOP Conf. Series: Materials Science and Engineering **714** (2020) 012003 doi:10.1088/1757-899X/714/1/012003

gradient images is developed. The technique allows to find the chain optimum on labor input in restrictions for the accuracy of recovery of circuits of the objects presented in the processed pictures. As an accuracy figure of recovery of circuits, the value of linear correlation coefficient between result of image processing by a chain of algorithms and the corresponding image template is used.

2.3.1 The enlarged technique stages. 1. Definition / replenishment of operations — groups of functional operators (algorithms) from which chains form: for each step of a solution of an objective set of alternative operators are defined. The following fact is very important: each group the "empty" operator surely is included, — not having a program code. "Inclusion" of this operator in a chain means that any of operators of this group is not used, i.e. the corresponding functional operation is not provided in the created algorithm.

As the initial rule set allowing to fill a pool of algorithms serves the system of axioms of Zermelo-Frenkel. If the offered algorithm when adding it to the existing system breaks this system of axioms, then the offered algorithm is discarded.

2. Determination of the reasonable sequence of operators defining structure of the possible chains forming an objective solution algorithm: a) alignment of the histogram of the image; b) filtering; c) selection of borders; d) specification of borders.

3. A task of value of the indicator characterizing the accuracy of recovery of a circuit of a product.

4. Implementation of search of an optimum chain by means of the special genetic algorithm as which fitness function the accuracy figure of recovery of a circuit of a product is used.

Generally for implementation of the genetic algorithm considerable time expenditure therefore it has special can be involved the number of attempts to find the best chain is exogenous the set λ-parameter.

5. Decision-making by results of work of the genetic algorithm is carried out: the "best" chain of the Pareto set created by the genetic algorithm is selected.

6. The report on that forms, — whether the chain of operators meeting the imposed precision requirements and if it is found is found, then data on structure of a chain, parameters of the operators entering this chain, both the accuracy figures and labor input provided with this chain are provided. If the chain providing the required accuracy of recovery of circuits is not found, then the fact of impossibility of a solution of an objective because of insufficiency of structure of the operators defined at the first stage, and need of return to the first stage for the purpose of replenishment of structure of functional operators is established.

The specified sequence of stages forms the methodical (enlarged) main flow of actions. Algorithmic the main and alternative flows are defined in the course of decomposition of the enlarged stages.

2.3.2 Specifics of the genetic algorithm applied to search of a chain of operators. First, feature of the search algorithm is that it carries out at the same time both parametrical, and structural identification of the created chains.

Secondly, its distinctive feature is the following: a) in the procedure of each mutation when forming the next chain of operators the sequence of operations is rigidly fixed, only the operators implementing operation are mutated; b) operation of alignment of the histogram of the image can be carried out only once; c) operation of filtering can repeat, at the same time both use of the same operator, and application of different operators of filtering is allowed; d) operation of selection of borders can be carried out only once; e) operation of specification of borders can be carried out only once. All operations, except alignment of the histogram, are obligatory.

Thirdly, specifics of the applied version of the genetic algorithm is that in the course of its work: a) not only fitness function (an accuracy figure of recovery of a circuit of a product), but also an indicator of labor input of the created chains of operators is estimated; b) the Pareto set with restriction, — containing data on the chains which are meeting the requirement to the accuracy of recovery of a circuit of a product, and, at the same time, not dominating each other on the accuracy and labor input forms.

Having executed forming of the next generation of chains, the main flow of the genetic algorithm transfers control to the flow implementing chains of the created generation which not only carries out operators of chains, but also estimates results of their work — the accuracy of recovery of a circuit and duration of work of all chains. These results are transmitted to a flow which conducts an ordered set of Pareto with restriction, using the approvals of the following theorems.

Theorem 1. The Pareto set with two criteria of quality of elements does not may contain elements with identical values of one-position criteria.

Theorem 2. Sorting of elements of a Pareto set with two criteria of quality of elements by one of criteria automatically leads to the inverse sorting according to the second criterion if criteria have the identical qualitative direction, and to direct sorting according to the second criterion if criteria are qualitatively multidirectional [9].

The stream of a Pareto set contains a set in the sorted state that allows to define quickly the potential place of a new line-up in this set, to find out a question of occurrence or not entry of a new line-up into a Pareto set and if necessary to delete the available chains dominated by a new line-up. Naturally, first, the question of admissibility of a new line-up becomes clear, — the fact of implementation of the requirement to accuracy of restitution of a contour of a product.

Upon completion of work of the main flow of the genetic algorithm the control is transferred to a method of making decision on the "best" chain of the created Pareto set.

Subjective representation: "Among private criteria there is one the most important, decisively defining quality of options", generates the next way of the choice.

1. Private criteria are ranged on the importance (importance) based on subjective representation of the decision-maker. For this purpose, can be used as methods of direct ranging (to criteria either ranks, or the points expressing their importance are directly assigned), and the methods of comparative analysis allowing to find ranks indirectly, — by results of paired comparisons of private criteria. Paired comparisons have strong preference when criteria can be organized in hierarchy.

2. The option, the first in most significant (main) criterion, appears the best.

Generally, the first in main for criterion can be several options. Therefore, the considered method should be added with the following item.

3. If by the main criterion the best are several options, then the lexicographical principle is used further: among results of the choice on the first criterion the choice by the second most important criterion is carried out; in need of use the subsequent criteria in decreasing order of their importance are involved.

Implementation of the lexicographic principle will surely come to the end with the choice of the only result, i.e. the choice by means of a random number generator will never be required. This fact is approved by the following theorem.

Theorem 3. The procedure of implementation of the lexicographical principle of the choice of the only best option in the conditions of multicriteria estimation of quality of options with number of private criteria certainly is productive it (is completed). At the same time the amount of the applied private criteria does not exceed $q-1$.

On the practical level the problem of the choice of option with two private criteria of quality is most relevant in problems of comparison of different algorithms of a solution of tasks: very often algorithms are compared by two qualitatively opposing criteria: accuracy and efficiency of a solution of some task. At the same time qualitatively positive criterion $K_{i,1}^{[+]}$ has semantics of accuracy of calculations, and qualitatively negative criterion $K_{i,1}^{[-]}$ has semantics of time of calculations. Multiplicative convolution of two private criteria in has an appearance:

$$K_i^{[\mathrm{M}2+]} = \frac{\bar{K}_{i,1}^{[+]} - \bar{K}_{1,\min}^{[+]}}{\bar{K}_{i,1}^{[-]} - \bar{K}_{1,\min}^{[-]}},\ i = 1, 2, \ldots, n. \qquad (2)$$

Semantic integrated criterion (2) represents gain of "productivity": it shows, — as far as the accuracy of calculations at increase by unit of time of calculations increases (in relation to the minimum value). The upper index [M2+] symbolizes multiplicativity, a two-criteria dimension and that fact that convolution is created for search of a maximum of gain of "productivity" (in numerator of a formula increment of qualitatively positive private criterion is placed).

On the basis of the specified convolution it is possible to offer the following rule of the choice of the best option has an appearance: as the best option by two qualitatively opposing criteria it is necessary to accept the option which is entering a Pareto set and having a maximum of relative advantage in this set. We will call the value of qualitatively positive criterion corresponding to a maximum of relative advantage a rational maximum, we will call the corresponding value of qualitatively negative criterion a rational minimum.

3. Calculation of an accuracy figure of recovery of circuits of the image

Let's designate as $\boldsymbol{P}$ a set of coordinates of pixels of the considered images, that is

$$\boldsymbol{P} = \{(x, y) \colon x \in \{0, 1, \ldots, M\},\ y \in \{0, 1, \ldots, N\}\},$$

where M, N — the number of pixels in images in vertical direction and horizontals respectively.

Let's call function $f(x, y)$ of the image the function displaying elements of $\boldsymbol{P}$ in values of intensity of a luminescence (brightness) of pixels of the image: $f : \boldsymbol{P} \to \boldsymbol{R}^n$ where $\boldsymbol{R}$ — a set of real numbers, n — the integral number determined by a format of representation of chromaticity of the image. For example, $n = 3$ for the RGB format, $n = 1$ for black-and-white images. We will limit further considerations to black-and-white images. Let's pay attention to the fact: when photographing function value of the image form by means of optoelectronic processes.

Let's define formally the image as a matrix of function values of the image: $\boldsymbol{F} = [f(x, y) : x \in (0, 1, \ldots, M),$ $y \in (0, 1, \ldots, N)]$.

Note: in literature, following [1], simply without the specifying formalization images call functions of the image, i.e. speak so: "let's consider images $f(x, y)$ and $g(x, y)$".

If the image $\boldsymbol{F}$ is exposed to processing by an algorithm $\boldsymbol{A}$ and as a result the new image $\boldsymbol{U} = \boldsymbol{A}(\boldsymbol{F})$ forms, then it is possible to write:

$$\boldsymbol{U} = [u(x, y) : x \in (0, 1, \ldots, M),\ y \in (0, 1, \ldots, N)],$$

where $u : \boldsymbol{P} \to \boldsymbol{R}$ — new function of the image.

If there is some additional image

$$\boldsymbol{V} = [v(x, y) : x \in (0, 1, \ldots, M),\ y \in (0, 1, \ldots, N)],$$

that it is possible to calculate linear correlation coefficient $r_{u,v}$ of images $\boldsymbol{U}$ and $\boldsymbol{V}$, determined by a formula:

$$r_{u,v} = \frac{\sum_{x=0}^{M-1}\sum_{y=0}^{N-1}\left[u(x, y) - \bar{u}\right]\left[v(x, y) - \bar{v}\right]}{\sqrt{\sum_{x=0}^{M-1}\sum_{y=0}^{N-1}\left[u(x, y) - \bar{u}\right]^2 \sum_{x=0}^{M-1}\sum_{y=0}^{N-1}\left[v(x, y) - \bar{v}\right]^2}}, \qquad (3)$$

where $u(x, y)$, $v(x, y)$ are values of functions of images, another words brightness of pixels with coordinates x, y; $\bar{u}, \bar{v}$ — mean values of brightness of pixels of images.

In the stated technique of synthesis of chains of the algorithms improving quality of gradient images, linear correlation coefficient $r_{u,v}$ between the gradient image $u(x, y)$ received as a result of processing of the picture of a product i -chain of algorithms, and the image template $v(x, y)$ considered as ideal result of processing of the same picture it is used as an accuracy figure of recovery of a circuit of the image. The coefficient $r_{u,v}$ expresses degree of similarity of images $u(x, y)$ and $v(x, y)$ therefore it is qualitatively positive criterion $K_i^{[1]}$ in the course of search of the best chain, other words $K_i^{[1]} = r_{u,v}$.

4. Calculation of indicators of labor input of functional operators and their chains

Estimation of indicators of labor input of functional operators by the classical technique [12, 13] of a pilot study based on calculation of number of the carried-out operations at different sets of input data is almost impossible because of high labor input of process of estimation and lack of source codes when using third-party programs. For this reason it is offered to estimate labor input on time expenditure.

As a labor input indicator d_i chain with index i of operators the sum of values of indicators of labor input of the functional operators who entered a chain $d_i = \sum_{j=1}^{n} d_{j_i}^{[j]}$, $i = \overline{1;m}$, where n — the number of functional groups of operators from whom chains form is used; m — quantity of the created chains; $d_{j_i}^{[j]}$ — an indicator of labor input of the operator functional group j, entered in chain i. The labor input indicator d_i-chains of operators i is qualitatively negative criterion $K_i^{[2]}$ in the course of search of the best chain: $K_i^{[2]} = d_i$.

5. Illustrative example of application of the offered technique for a solution of an applied task

For a solution of a specific applied objective on improvement of gradient images of products of aerospace or railway assignment in the course of control of their overall dimensions [10] it was enough to use a small number of algorithms (see the table). This set allows to eliminate the rejections of the processed image from a standard caused by the different reasons including:

1) the greasing caused by the movement of the conveyer belt;

2) change of illumination;

3) the accidental beaten-out pixels on the image which arose because of defect of a matrix of the camera (accidental hit of dust, purity of materials and quality of a substrate of a matrix).

Application of a threshold in algorithms of selection of borders allows to consider pixel belonging to border if the pixel of result of filtering corresponding to it matters, bigger the specified value. Feature of implementation of the operator Canny in the environment of Matlab is the possibility of application of the threshold represented by either a scalar variable, or a two-element vector. For classification of differences on "weak" and "strong" two thresholds — lower and upper are used. "Weak" borders are noted in the resulting image, only if they are connected to "strong". If parameter a threshold is scalar value, then the threshold sets value of an upper limit, and for a lower limit the value 0.4*threshold is used. Input parameter, depending on type of submission of the image can be both integer, and material. The offered modification of the genetic algorithm is initially configured for work with material values.

Parameter the direction is used only in Sobel's operator and can accept one of three values:

- 'horizontal' — selection of horizontal borders;
- 'vertical' — selection of vertical borders;
- 'both' — selection of borders in all directions.

Selection of borders only of one look is unreasonable therefore in the offered modification of the genetic algorithm search of borders in all directions is used.

The images received by photography of products under production conditions were rather good, and the program implementing the offered technique found a chain of the operators providing correlation coefficient between result of processing and a template not less $r_{u,v}^{[\min]} = 0.97$ at the assigned permissible value of coefficient $r_{u,v}^{[\text{add}]} = 0.95$. Time expenditure at the same time made from 1 to 100 ms. $r_{u,v}^{[\min]} = 0.97$.

For observation of behavior of a technique in more difficult conditions the experiment with artificially impaired a little image was made — on the images received by photography ten-percentage was imposed noise is Gaussian (by installation of the corresponding value of the effect parameter of a packet Adobe Photoshop). One of test images is given in figure 1.

During a solution the genetic algorithm creates also impractical individuals, — with the values of parameters which are not making physical sense or overstepping the bounds of rational intervals.

In the course of search of the best chain of algorithms the share of impractical individuals made from 15 to 18%. Impractical individuals appeared in each generation of population.

Table 1. Applied image processing algorithms, their parameters and features.

Algorithm	Parameters	Features
Filtering algorithm		
«Empty statement»	—	There is no effect on the image. Included in the list to optimize the time of the algorithm: used if the original image does not need filtering
Averaging filter	Radius	The radius is given by a two-component vector
Gauss filter	Radius, standard deviation	The standard deviation parameter of the Gaussian distribution, which is used when forming the mask
Median filter	Radius	The slowest filtering algorithm. The filter does not distort the sharp edges of the object
Edge detection operators		
Sobel operator	Threshold, direction	A pixel is considered to belong to a border if the corresponding pixel of the filtering result has a value greater than threshold
Prewitt operator	Threshold	
Canny operator	Threshold, standard deviation	
Laplacian – Gaussian operator	Threshold, standard deviation	The boundaries are selected after applying a filter similar to the sequential application of the Gauss and Laplace filters
Roberts operator	Threshold	A pixel is considered to be a boundary if its corresponding filtering result pixel has a value greater than threshold

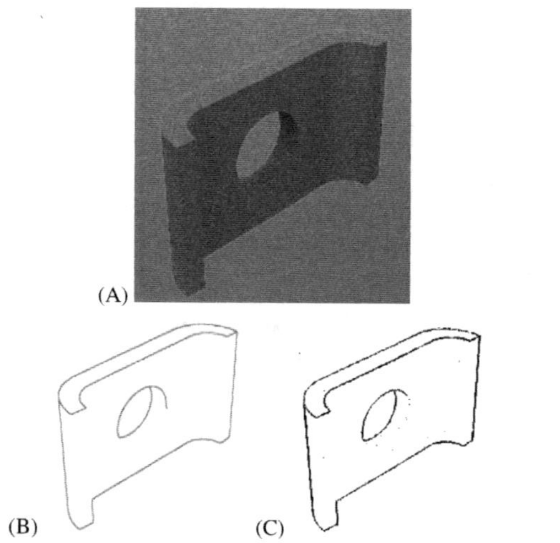

Figure 1. (A) test image of a clamping bracket; (B) the image standard (circuits are selected by the expert); (C) result of processing by the chain generated by the genetic algorithm.

About a half of viable individuals mattered fitness function (linear correlation coefficient $r_{u,v}$), exceeding 0.75. Quality characteristics of such individuals were used for forming of statistical data.

For obtaining required result search was limited to seven generations. The received final Pareto set had power from 47 to 573. The chains qualified as the first in criteria of a maximum of relative advantage mattered linear correlation coefficient (a rational maximum of qualitatively positive criterion $K_i^{[1]} = r_{u,v}$) from 0.95 to 0.98. At the same time values of a rational minimum of qualitatively negative criterion $K_i^{[2]} = d_i$ (duration of the best chains) were in the range from 8 up to 110 ms.

The diagram of dependence of accuracy of calculations from a counting duration of fifteen algorithms synthesized by means of the genetic algorithm [11], which entered a dynamic Pareto set is provided on figure 2.

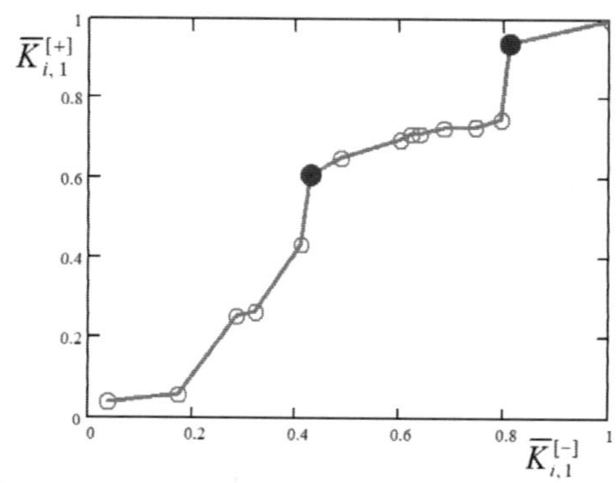

Figure 2. Dependence of accuracy of calculations on time expenditure.

Circles on the diagram correspond to the algorithms which entered a Pareto set. On abscissa axis the normalized algorithm operating time — the real time divided into the maximum value of a counting duration among the algorithms which entered a Pareto set is postponed. On ordinate axis the normalized algorithm accuracy (correlation coefficient) — the real coefficient divided into the maximum value of coefficient among the algorithms which entered a Pareto set is postponed.

Vector of values of integrated criterion (2):

$K_i^{[M2+]} = [1; 0.1387; 0.8608; 0.7931; 1.053; 1.45; 1.355; 1.164; 1.141; 1.114; 1.059; 0.9733; 0.9319; 1.164; 0.9947]$.

Graphically each value of a vector $K_i^{[M2+]}$ is a tangent of angle of an inclination to abscissa axis of the straight line connecting a starting and current point of the diagram.

Considering values of vector $K_i^{[M2+]}$, it is possible to establish that according to criterion (2) the best is the sixth algorithm.

6. Conclusions

The main importance of work consists that in it the new technique of completely automatic synthesis of chains of the algorithms improving quality of gradient images is stated. The following new and original ideas are the cornerstone of the developed technique.

1. Use of a formalism — the system of axioms of Zermelo-Frenkel at the initial stage of forming of groups of functional operators (algorithms) from which chains form; each functional group the empty operator is included; the specified innovations allow to carry out correct forming of functional groups in the automatic mode.

2. Automatic simultaneous structural and parametrical identification of a synthesizable chain of operators.

3. Application as the mechanism of implementation automatic identification of a synthesizable chain of operators of the modified genetic algorithm. Modification consists in combination of the procedure of search with simultaneous maintaining a dynamic Pareto set and provides an opportunity "industrial" (without participation of the expert) uses of the offered technique.

4. Use of two criteria of quality of a synthesizable chain of operators: 1) accuracy of recovery of a circuit of a product; 2) minimum of labor input of implementable operations (image processing time chain).

5. Application of a new technique of the two-criteria optimum choice of a solution for tasks with dynamic generation of options, with in advance uncertain number of options. The technique is created for implementation of a two-criteria-process of synthesis of a chain of operators. It is independently significant, — can be used in different applications of the two-criteria optimum choice.

Theoretical results of work are:

1) a technique of completely automatic synthesis of chains of the algorithms improving quality of gradient images;

2) the modified genetic algorithm combining the procedure of search with simultaneous maintaining a dynamic Pareto set;

3) a technique of the two-criteria optimum choice of a solution for tasks with dynamic generation of options;

its conceptual elements:

- criterion of a maximum of relative advantage;
- theorems of a Pareto set with two quality characteristics of elements;
- concepts of a rational maximum and minimum of two qualitatively multidirectional criteria;
- restrictions for criteria;

its procedural elements:

- generation of options;
- maintaining dynamic Pareto set;
- decision-making by criterion of a maximum of relative advantage.

The created technique of automatic synthesis of chains of algorithms and contextually developed independently significant technique of the two-criteria optimum choice of a solution have both theoretical, and practical value as fill up structure of methodical ensuring practice of a solution of the tasks connected as with image processing, and acceptance of optimal solutions.

References

[1] Gonzalez R C and Woods R E 2018 *Digital Image Processing* (Pearson Prentice Hall)

[2] Munteanu C and Rosa A 2004 Gray-Scale Image Enhancement as an Automatic Process Driven by Evolution *IEEE Transactions on Systems, Man, and Cybernetics* **34** 1292

[3] Makhno T A Automatic construction of combinations of image processing algorithms in the problem of processing ultrasound images of carotid arteries *Graphicon 2014 Proceedings* pp 104–7 https://www.graphicon.ru/html/2014/papers/104–107.pdf

[4] Makhno T A Automated system for processing ultrasound images of carotid arteries based on evolutionary algorithms http://abstracts.donnu.edu.ua/article/view/2465

[5] Belousov A A, Spicyn V G and Sidorov D V 2006 *Izvestiya Tomskogo politekhnicheskogo universiteta* **7** 21

[6] Chernyavskii A B and Spicyn V G 2006 *Izvestiya Tomskogo politekhnicheskogo universiteta* **7** 26

[7] Belousov A A and Spicyn V G 2009 The two-stage method to improve the images *Tekhnologii Microsoft v teorii i praktike programmirovaniya* pp 123–5

[8] Teryokhin A V 2015 Three-dimensional object recognition system https://dlib.rsl.ru/01008053130

[9] Below V V and Lopatin A K 2018 *Vestnik Ryazanskogo gosudarstvennogo radiotekhnicheskogo universiteta* **66** 78

[10] Kuznecov A E and Ryzhikov A S 2014 *Vestnik Ryazanskogo gosudarstvennogo radiotekhnicheskogo universiteta* **48** 11

[11] Belov V V and Lopatin A K 2016 *XXI vek: itogi proshlogo i problemy nastoyashchego plyus* **6** 34

[12] Petrushin V N and Ul'yanov M V 2008 *Informacionnye tekhnologii i vychislitel'nye sistemy* **2** 81

[13] Cormen T H, Leiserson C E, Rivest R L and Stein C 2016 *Introduction to Algorithms* (The MIT Press Cambridge Massachusett)

Nonparametric Method for Aircraft Sensor Fault Real-Time Detection and Localization

Y Bondarenko[1], A Chekin[1], E Zybin[1] and V Kosyanchuk[1]

[1] State Research Institute of Aviation Systems, Moscow, Russia

E-mail: zybin@mail.ru

Received xxxxxx
Accepted for publication xxxxxx
Published xxxxxx

Abstract

This paper is a part of a series of articles on unified nonparametric methods in dynamic systems theory. Here the authors propose a new method for dynamic system sensor fault real-time detection and localization based only on some past measurements of monitored system input-output signals. The described nonparametric method needs no any priori information on system or sensor model parameters and does not require functional redundancy, identification, prediction, training or statistical calculations. An example of nonparametric detection and localization of aircraft multiple sensor faults is presented.

Keywords: aircraft, sensor, fault, detection, localization, nonparametric, model-free, real-time, parametric uncertainty

1. Introduction

There are a large number of methods for dynamic system sensor fault detection and localization [1–20]. All of them can be divided into two broad groups: parametric (model-based) [8–14] and nonparametric (model-free, data-driven, data-based, signal-based, history-based, etc.) [15–20].

Parametric (model-based) methods are the best known and are considered classical. They directly or indirectly use the parameters of mathematical models of real systems, the values of which are given a priori or evaluated during identification. Two approaches to sensor fault monitoring by model-based methods in the parameter and signal spaces are known.

In the first case, the current values of some system parameters, which are compared with their nominal values, are estimated using parametric identification methods. The main limitation of this approach in aircraft sensor monitoring is the impossibility of identifying the open-loop model parameters of an aircraft under closed-loop control [21]. To ensure the aircraft model identifiability in such cases, it is necessary to apply active identification algorithms with feeding the high-excited signals directly to the control actuators. This can significantly reduce the aircraft flight safety, especially in the events of sensor fault.

The signal space model-based methods are focused on generating and analyzing some residual between the output signals of a real system and its mathematical model. The most common way to generate such a residual is to restore and predict the system state vector with the subsequent determination of its measurement vector. However, to solve these problems it is also necessary to know the exact and actual values of system model parameters, that can be difficult in practice [22–24]. The inevitable model errors lead to an increase in the monitoring criteria threshold values, which, in turn, leads to an increase in the faults detection time and to a decrease in localization reliability.

Nonparametric (model-free) methods, unlike traditional ones, do not require information about the parameters of monitored system models and are based only on the analysis of measurements of their input and output signals. Such methods consider the monitored system in the form of a "black box" and allow monitoring problem solving under complete parametric uncertainty. The well-known nonparametric methods require either their preliminary training or long-time tuning for a specific system (neural network, genetic, fuzzy logic, etc.), or based on statistical (probabilistic) algorithms. All statistical algorithms require a sufficiently large data sample for the estimating of the statistical properties of the analyzed variables, which inevitably leads to an increase in the fault detection and localization time. In flight, the application of such approaches may require a time exceeding the control system critical response time, and the aircraft may go into an unrecoverable state.

This paper extends application area of a series of previously developed unified nonparametric methods in dynamic systems theory on sensor fault detection and localization problems [25–30]. The proposed method is based on the measurements of the aircraft controls and states only. It doesn't require a priori information of the aircraft model parameters, statistical calculations, long time tuning or training. It has a very simple form and is based on algebraic solvability condition for the aircraft model identification equation. The scope of the paper is limited to deterministic stationary discrete-state linear completely observable models.

2. Nonparametric aircraft sensor fault detection and localization problem

Let the models of aircraft dynamics with non-faulted and faulted sensors are represented in the state space as

$$x_{i+1} = Ax_i + Bu_i,\ y_i = Cx_i, \tag{1}$$

$$x_{j+1} = Ax_j + Bu_j,\ y_j = FCx_j, \tag{2}$$

where A, B, C — the eigen-dynamics, control efficiency and measurement matrices; x, u, y — the state, control and measurement vectors of length n_x, n_u, n_x; $i = 0,1,...,l-1$; $j = l, l+1,...$ — the discrete times before and after the occurrence of sensor fault; l — the instant a fault occur; $F = \operatorname{diag}\left[f(1) \ \cdots \ f(k) \ \cdots \ f(n_x)\right]$ — the sensor fault (calibration error) matrix, $f(*) = 1$ for a non-faulted sensors, $0 < f(*) < 1$ for faulted sensors.

It is necessary, without any priori information on the aircraft model parameters *A, B, C,* based on the controls *u* and measurements *y* only, to detect the time of fault and to localize the faulted aircraft sensors.

Suppose that we have some series of the measurements. Then the aircraft models (1), (2) can be written in matrix forms

$$X_{i+1} = AX_i + BU_i,\ Y_i = CX_i, \tag{3}$$

$$X_{j+1} = AX_j + BU_j,\ Y_j = FCX_j, \tag{4}$$

where h, h^f — the numbers of measurements in non-faulted and faulted cases, $X_{i+1} = \left[x_{i+1} \ x_{i+2} \ \ldots \ x_{i+h}\right]$, $X_i = \left[x_i \ x_{i+1} \ \ldots \ x_{i+h-1}\right]$, $U_i = \left[u_i \ u_{i+1} \ \ldots \ u_{i+h-1}\right]$, $Y_i = \left[y_i \ y_{i+1} \ \ldots \ y_{i+h-1}\right]$.

For completely observable model ($C^{-1}C=I$) the state matrices can be recovered by the expressions

$$X_i = C^{-1}Y_i, \tag{5}$$

$$X_j = C^{-1}F^{-1}Y_j. \tag{6}$$

Then, substituting (5), (6) to (3), (4) gives us the equivalent measurement-space models

$$Y_{i+1} = CAC^{-1}Y_i + CBU_i,$$

$$Y_{j+1} = FCAC^{-1}F^{-1}Y_i + FCBU_i,$$

which can be represented in the form of the matrix equations

$$\begin{bmatrix} CAC^{-1} & CB \end{bmatrix} \begin{bmatrix} Y_i \\ U_i \end{bmatrix} = Y_{i+1}, \tag{7}$$

$$\begin{bmatrix} FCAC^{-1}F^{-1} & FCB \end{bmatrix} \begin{bmatrix} Y_j \\ U_j \end{bmatrix} = Y_{j+1}. \tag{8}$$

It is known, that any linear matrix equation of the form $ZQ = W$ with known matrices *Q*, *W* is solvable for *Z* if and only if the necessary and sufficient solvability condition is satisfied [25]

$$WQ_R^{\perp} = 0, \tag{9}$$

where $Q_R^{\perp}$ — the right-hand full rank zero divisor, such that $QQ_R^{\perp} = 0$.

According to (9) the equations (7), (8) are solvable if and only if the following conditions are satisfied

$$Y_{i+1}\begin{bmatrix} Y_i \\ U_i \end{bmatrix}_R^{\perp} = 0,\ Y_{j+1}\begin{bmatrix} Y_j \\ U_j \end{bmatrix}_R^{\perp} = 0, \tag{10}$$

where

$$\begin{bmatrix} Y_i \\ U_i \end{bmatrix}\begin{bmatrix} Y_i \\ U_i \end{bmatrix}_R^{\perp} = 0,\ \begin{bmatrix} Y_j \\ U_j \end{bmatrix}_R^{\perp}\begin{bmatrix} Y_j \\ U_j \end{bmatrix}_R^{\perp} = 0. \tag{11}$$

But when the sample window includes both non-faulted and faulted measurements there is no any linear model corresponding with measured signals

$$\begin{bmatrix} Y_{i+1} & Y_{j+1} \end{bmatrix} \neq \hat{A}\begin{bmatrix} Y_i & Y_j \end{bmatrix} + \hat{B}\begin{bmatrix} U_i & U_j \end{bmatrix},$$

and the solvability condition does not hold

$$\begin{bmatrix} Y_{i+1} & Y_{j+1} \end{bmatrix}\begin{bmatrix} U_i & U_j \\ Y_i & Y_j \end{bmatrix}_R^{\perp} \neq 0. \tag{12}$$

Having viewed (12) line by line, it can be seen that for the non-faulted measurement channels the solvability conditions are still satisfied, because they are generated by original non-faulted linear model. This allows us to use the row norms of (12) for the k of n_x measurement channels $Y_* = \left[y_*^{(1)} \ldots y_*^{(k)} \ldots y_*^{(n_x)}\right]^{\mathrm{T}}$

$$\sigma^{(k)} = \left\| \begin{bmatrix} y_{i+1}^{(k)} & y_{j+1}^{(k)} \end{bmatrix}\begin{bmatrix} U_i & U_j \\ Y_i & Y_j \end{bmatrix}_R^{\perp} \right\|_2 \neq 0 \tag{13}$$

as nonparametric criteria for aircraft sensor fault detection and localization.

3. Nonparametric aircraft sensor fault detection and localization example

To verify the proposed method a midrange passenger aircraft dynamics was simulated by the discrete state-space model (1), where [25]:

$$A=\begin{bmatrix} A_{11} & 0 \\ 0 & A_{22} \end{bmatrix},\ B=\begin{bmatrix} B_{11} \\ B_{21} \end{bmatrix},\ C=I,$$

$$A_{11}=\begin{bmatrix} 0,9998 & -0,1021 & -0,0981 & 0,0000 \\ -0,0001 & 0,9907 & 0,0000 & 0,0100 \\ 0,0001 & 0,0093 & 0,9999 & 0,0000 \\ 0,0000 & -0,0079 & 0,0000 & 0,9973 \end{bmatrix},$$

$$A_{22}=\begin{bmatrix} 1,0003 & 0,0011 & 0,0099 & 0,0014 \\ -0,0344 & 0,9650 & -0,0090 & 0,0000 \\ -0,0199 & 0,0020 & 0,9896 & 0,0000 \\ 0,0000 & 0,0100 & -0,0011 & 1,0000 \end{bmatrix},$$

$$B_{11}=\begin{bmatrix} 0 & 0 & 0 & 0 \\ 0.00018 & 0.00018 & -0.00008 & 0 \\ -0.00018 & -0.00018 & 0.00008 & 0 \\ -0.00338 & -0.00338 & -0.01343 & 0 \\ 0 & 0 & 0 & -0.00005 \\ 0.00205 & -0.00205 & 0 & -0.00197 \\ -0.00009 & 0.00009 & 0 & -0.03146 \\ 0 & 0 & 0 & 0 \end{bmatrix},$$

$$B_{21}=\begin{bmatrix} 0 & 0 & 0.00005 & 0.00005 \\ -0.00001 & -0.00001 & -0.00007 & -0.00007 \\ 0.00001 & 0.00001 & 0.00007 & 0.00007 \\ -0.0005 & -0.0005 & 0.00127 & 0.00127 \\ 0 & 0 & -0.00002 & 0.00002 \\ 0.00542 & -0.00542 & -0.01636 & 0.01636 \\ -0.00058 & 0.00058 & 0.00297 & -0.00297 \\ 0 & 0 & 0 & 0 \end{bmatrix},$$

$x=[\Delta V \ \ \Delta\alpha \ \ \Delta\vartheta \ \ \Delta\omega_z \ \ \Delta\beta \ \ \Delta\omega_x \ \ \Delta\omega_y \ \ \Delta\gamma]^{\mathrm{T}}$ — the state vector of flying speed (km/h), angles of attack, pitch, slip, and roll (deg), angular rates of roll, yaw, and pitch (deg/s); $u=[u_{\mathrm{e\_l}} \ \ u_{\mathrm{e\_r}} \ \ u_{\mathrm{st}} \ \ u_{\mathrm{ru}} \ \ u_{\mathrm{a\_l}} \ \ u_{\mathrm{a\_r}} \ \ u_{\mathrm{i\_l}} \ \ u_{\mathrm{i\_r}}]^T$ — the control vector of right and left elevators, stabilizer, rudder, right and left ailerons, right and left interceptors deflection angles (deg).

Fig. 1 shows the modelled aircraft controls of the form sin(wt) with periodic deflection of the rudder (w=0.1 rad/sec), in-phase deflection of elevators (w=0.05 rad/sec) and differential deflection of ailerons (w=0.007 rad/sec). The stabilizer and interceptors were set in zero positions.

The single faults $f(*)$=0.9999 were sequentially injected every 5 seconds in every sensor and the multiply fault with the same calibration errors in all rotational speed sensors was injected at 45 second.

Fig. 2 shows the state measurements for non-faulted and faulted cases. As one can see, such kind of faults are visually undetectable at this scale because the graphs in non-faulted and faulted cases are fully coincide.

Fig. 3, 4 show the values of detecting and localizing criteria (13) for sample width h=17 in different time scales. As one can see, they are close to zero within the accuracy to computational errors when the aircraft model parameters remain constant. All sensor faults result in the clearly visible pulses with amplitudes of 4–6 orders and widths equal to h+1. These results confirm the high sensitivity of proposed method concerning sensor calibration error in fourth decimal place only.

From fig. 4 it is obvious that the algorithm can detect and localize sensor fault immediately after the first faulted measurement becomes available and has a very short tune time. For this example with sample rate $\Delta t = 0.01$ seconds and the nonparametric algorithm has tune time $t_{tune} = \Delta t(h+1) = 0.18$ seconds only. This is due to the fact that in the current situation for the existence of the right-hand zero divisors (11) it is sufficient to accumulate $n_x+n_u+1=17$ columns of past states and controls to detect their linear dependences.

4. Conclusion

As a result the new nonparametric method for aircraft sensor fault real-time detection and localization, based only on control and state measurements, is developed.

The main advantage of the method is its independence from the aircraft model parameters, that completely eliminates modelling errors and guarantees its efficiency in a full parametric uncertainty even in the case of aircraft model unidentifiability. It does not require sensor functional redundancy, test control signals, or state observing, model identification and aircraft dynamic prediction problems solution.

Unlike wide known nonparametric methods it has a very simple form and need no a priory training, long time tuning or statistical calculations. This makes it possible to significantly increase the reliability and speed of aircraft sensor fault real-time detection and localization problem solution.

Acknowledgments

This work was supported by Russian Foundation for Basic Research (RFBR), projects 18-08-00453a, 20-08-01215a, 19-29-06091mk.

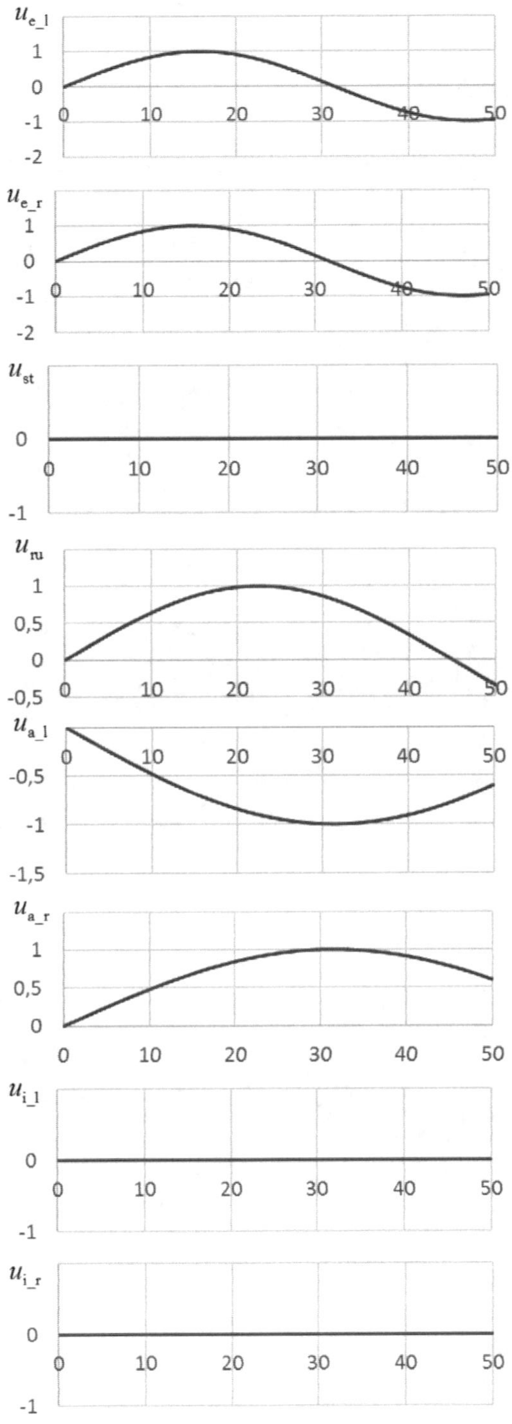

Figure 1. Aircraft controls.

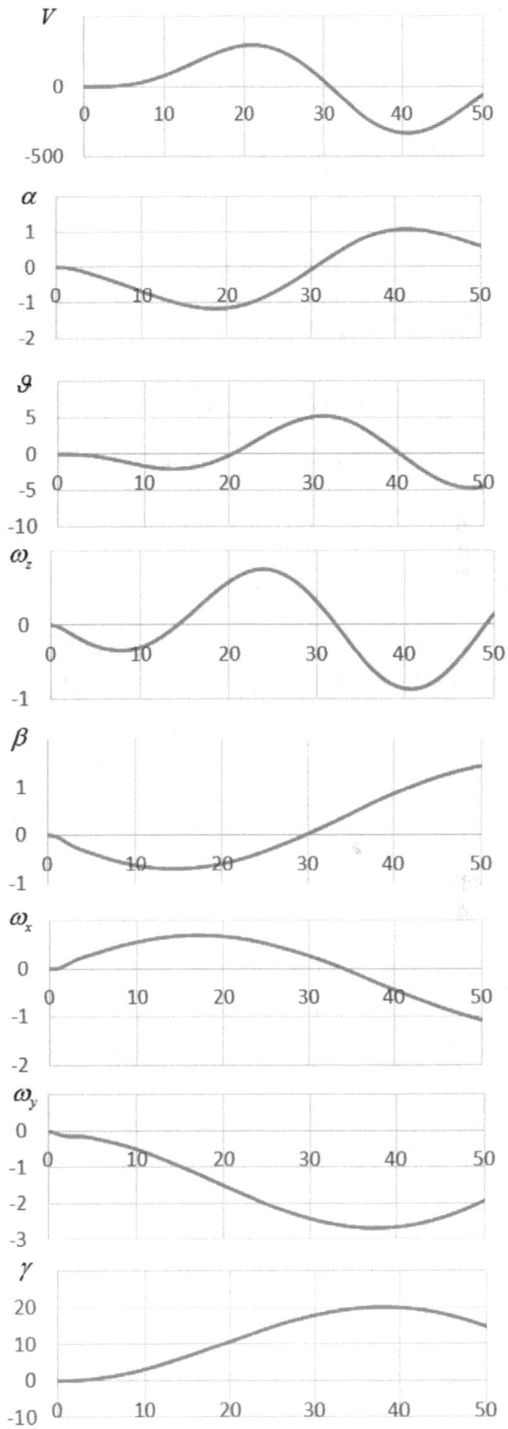

Figure 2. Aircraft state measurements.

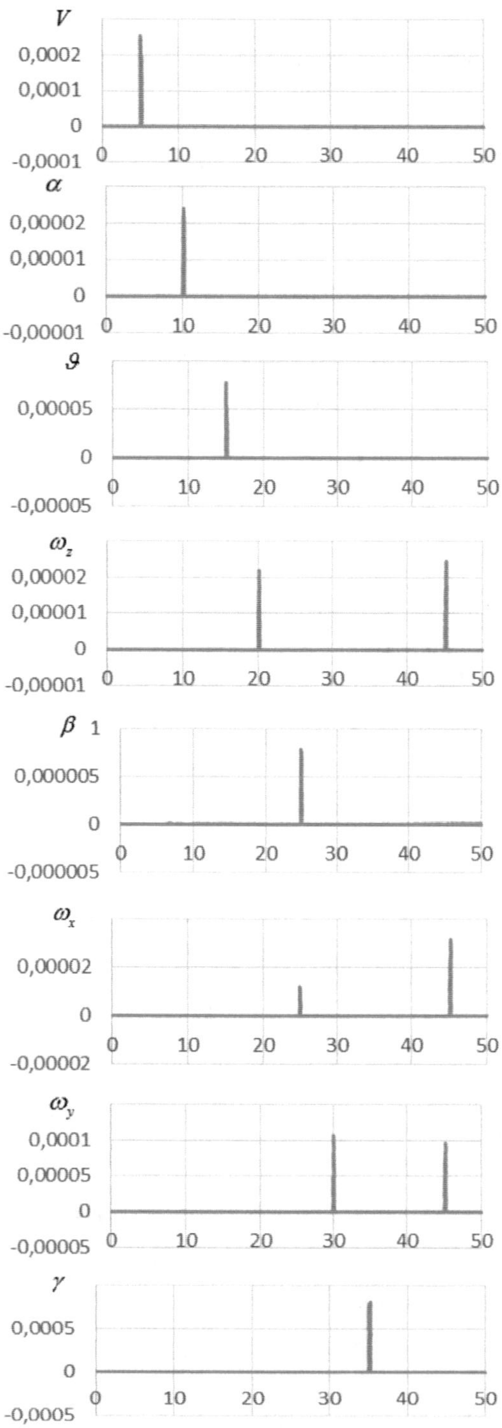

Figure 3. Sensor fault detection and localization criteria.

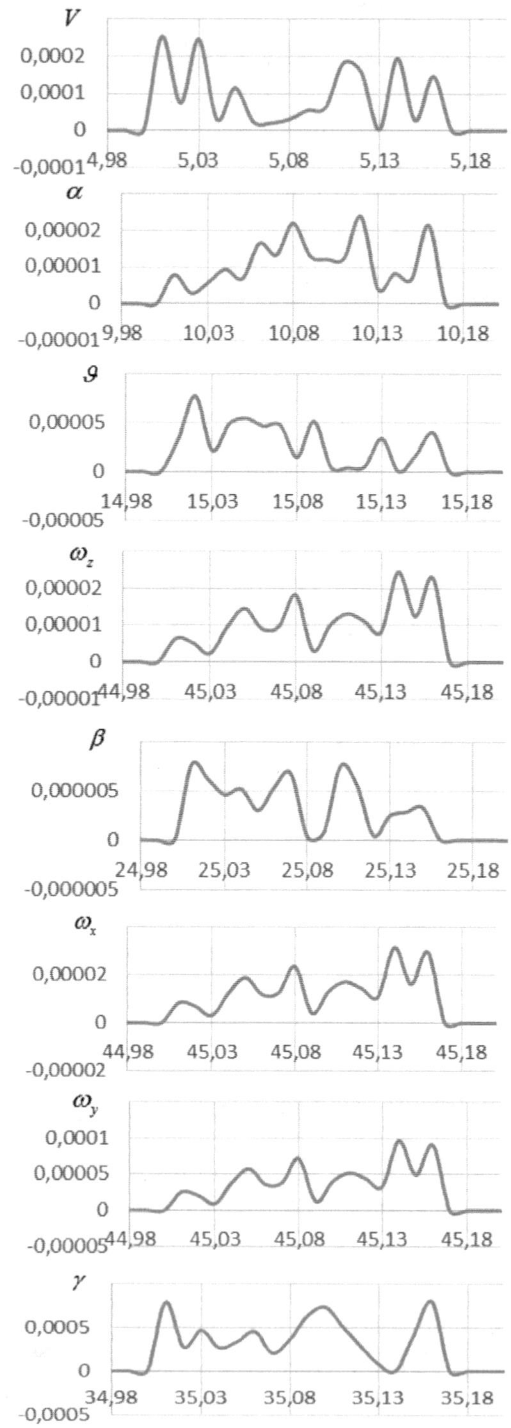

Figure 4. Sensor fault detection and localization scaled-up criteria.

References

[1] *Zolghadri* A, Henry D, Cieslak J, Efimov D and Goupil P 2013 *Fault diagnosis and fault-tolerant control and guidance for aerospace vehicles: from theory to application* (London: Springer)

[2] Hajiyev C and Caliskan F 2013 *Fault diagnosis and reconfiguration in flight control systems* (NY: Springer)

[3] Qi X, Theilliol D, Qi J, Zhang Y, Han J, Song D, Wang L and Xia Y 2013 *Conference on Control and Fault-Tolerant Systems (SysTol)* pp 132–9

[4] Hou Z S and Wang Z 2013 *Information Sciences* **235** 3–35

[5] Samy I, Postlethwaite I and Gu D W 2011 *Control Engineering Practice* **7** 658–74

[6] Heredia G, Ollero A Bejar M and Mahtani R 2008 *Mechatronics* **2** 90–9

[7] Bonfe M, Castaldi P, Geri W and Simani S 2006 *International journal of adaptive control and signal processing* **8** 381–408

[8] Ding S 2012 *Model-based fault diagnosis techniques: design schemes, algorithms and tools* (London: Springer)

[9] Fravolini M L, Napolitano M R, Del Core G and Papa U 2018 *Control Engineering Practice* **78** 196–212

[10] De Loza A F, Cieslak J, Henry D, Dávila J and Zolghadri A 2015 *IET Control Theory & Applications* **4** 598–607

[11] Ansari A and Bernstein D S 2016 *American Control Conference (ACC)* 5951–6

[12] Lu P, Van Eykeren L, Van Kampen E J and Chu Q P 2015 *AIAA Guidance, Navigation, and Control Conference* 1311

[13] Freeman P, Seiler P and Balas G J 2013 *Control Engineering Practice* **10** 1290–301

[14] Eubank R D, Atkins E M and Ogura S 2010 *AIAA Guidance, Navigation and Control Conference* 1–14

[15] Linares R, Vittaldev V and Godinez H C 2018 *Handbook of Dynamic Data Driven Applications Systems* (Cham: Springer)

[16] El-Koujok M, Benammar M, Meskin N, Al-Naemi M and Langari R 2014 *Information Sciences* **259** 346–58

[17] Fravolini M L, Del Core G, Papa U, Valigi P and Napolitano M R 2019 *IEEE Transactions on Control Systems Technology* **1** 234–48

[18] Hou Z and Jin S 2013 *Model free adaptive control: theory and applications* (NW: CRC press)

[19] Sarkar S, Jin X and Ray A 2011 *Journal of Engineering for Gas Turbines and Power* **8** 081602

[20] Wang K, Chen J and Song Z 2017 *Journal of Process Control* **54** 152–71

[21] Zybin E Yu 2015 *Izvestiya SFedU. Engineering Sciences* **4** 160–70

[22] Korsun O N, Stulovskii A V, Balyk O A and Zolotaeva M V 2018 *IOP Conference Series: Materials Science and Engineering* **312** 012015

[23] Korsun O N, Stulovskii A V, Ovcharenko V N and Kanyshev A V 2018 *Journal of Computer and Systems Sciences International* **3** 374–89

[24] Korsun O N, Om M H, Latt K Z and Stulovskii A V 2017 *Procedia Computer Science* 67–74

[25] Chekin A Yu, Bondarenko Yu V, Zybin E Yu and Kiselev M A 2019 *IOP Conference Series: Materials Science and Engineering* **476** 012003

[26] Glasov V V, Zybin E Yu and Kosyanchuk V V 2019 *IOP Conference Series: Materials Science and Engineering* **476** 012011

[27] Karpenko S S, Zybin E Yu and Kosyanchuk V V 2018 *IOP Conference Series: Materials Science and Engineering* **312** 012010

[28] Terent'yev M N, Karpenko S S, Zybin E Yu and Kosyanchuk V V 2018 *IOP Conference Series: Materials Science and Engineering* **312** 012025

[29] Zybin E, Kosyanchuk V and Karpenko S 2017 *MATEC Web of Conferences* **99** 03011

[30] Zybin E Y and Kos'yanchuk V V 2016 *Journal of Computer and Systems Sciences International* **4** 546–57

Recurrent Neural Networks' Configurations in the Predictive Maintenance Problems

L A Demidova[1]

[1] MIREA Russian Technological University, Moscow, Russia

E-mail: demidova.liliya@gmail.com

Received xxxxxx
Accepted for publication xxxxxx
Published xxxxxx

Abstract

The possibilities of various configurations of the recurrent neural networks in solving the problems of the maintenance performance based on the multidimensional time series have been investigated. The typical examples of the maintenance performance' problems from technical and medical diagnostics have been considered. The configurations' examples of the one- and two-layer recurrent neural networks with the RNN, LSTM, and GRU neurons for the aircraft engine maintenance problems have been given, the graphical dependencies of the development' results of the neural network models, the estimates of the development time, and the estimates of the accuracy indicator have been presented. The conclusions about the advantages of the recurrent neural networks with the LSTM and GRU neurons have been made.

Keywords: predictive maintenance, deep learning, machine learning, recurrent neural network, long short term memory, gated recurrent unit, neural network structure

1. Introduction

The problem of forecasting and classifying the state of objects is relevant in various application areas. A herewith, the state of object at any time can be described using the readings of the large number of sensors. Therefore, such problem can be considered as a non-destructive testing problem and is the problem of analyzing multidimensional time series. A solution to this problem can be obtained using the modern data mining tools and the machine learning technologies, for example, using the classifiers and regression models based on the SVM algorithm (Support Vector Machine algorithm) [1–3], RF algorithm (Random Forest algorithm) [4, 5], artificial neural networks [6–9]. In recent years, the deep learning technologies based on the neural networks [9] have been actively used in solving the problems of forecasting and classifying the state of objects on the basis of the large number of sensors.

Predictive maintenance is a domain where data is collected over time to monitor the state of the object with the goal of finding the patterns to predict the failures [10]. Predictive maintenance includes a variety of aspects, for example: failure prediction, failure diagnosis (root cause analysis), failure detection, failure type classification, and recommendation of mitigation or maintenance actions after failure.

The effective solution to the problems of the proactive intellectual maintenance can be obtained precisely with the use of the deep learning technologies.

Among the deep learning technologies, LSTM (Long Short Term Memory) and GRU (Gated Recurrent Unit) networks, which are essentially recurrent networks (RNN), are especially appealing to the predictive maintenance domain due to the fact that they are very good at learning from sequences [11, 12]. This fact allows the use of time series data in the development of the neural networks, operating with longer periods of time from the past to detect the failure patterns.

2. Formulation of the typical problems

The typical examples of the proactive intelligent maintenance problems are the following problems, mentioned below. The first two of these problems belong to the sphere of technical diagnostics, the third problem belongs to the sphere of medical diagnostics.

The first problem can be formulated with the aim to predict the remaining useful life of aircraft engines. The created template must use the aircraft sensors' values to predict when an aircraft engine will fail in the future so that maintenance can be planned in advance [13–15].

The experimental dataset, which can be used to test new methods of maintenance in this sphere, contains the engine id for 100 engines, the cycle number of a specific engine (a herewith, the number of the last cycle is the point of failure of stable operation of the equipment); the initial wear and production differences (3 parameters in total), the readings of 23 sensors during the engine operation over the operation cycle. A herewith, each engine can have its own last cycle describing the point of failure of stable operation of the equipment.

The second problem can be formulated with the aim to propose the early failure detection methods for hard disk drives in order to improve storage systems availability and avoid data loss [16–18]. Failure prediction in such circumstances would allow for the reduction of downtime costs through anticipated disk replacements. Hard disk failure prediction plays an important role in reducing equipment downtime and improving service reliability. Many of the known methods, which applied for this problem, consider it as the classification problem and perform the incipient failure detection thus not allowing for proper planning of the maintenance problems. Others perform well only under a limited prediction horizon. It is actually to develop the methods, which can be capable to directly predict the remaining useful life (RUL) estimation for the hard disk drives based on SMART parameters. These methods must be able to predict the failures in both long and short term intervals by leveraging the capabilities of the LSTM and GRU neurons.

The experimental dataset, which can be used to test new methods of maintenance in this sphere, is comprised of the daily observation of 92,348 hard disk drives (HDD) during some years. These observations contain information regarding the serial number, model, capacity, fault, and 90 SMART attributes of each device. According to the Backblaze Company, a device is labeled as faulty if it stops working (does not turn on or does not receive commands), or if SMART self-test fails for attributes 5, 187, 188, 197, or 198. Most HDD models do not report all SMART attributes. In this case, the values not reported are left blank. In addition, different manufacturers and device models may report different attributes. All this information is consolidated in the experimental dataset.

The third problem can be formulated with the aim to propose types of therapeutic intervention for patients [19–21]. The heterogeneity among patients with various diseases usually leads to the different progression patterns and may required different types of therapeutic intervention. Hence, it is important to study patient subtyping, which is grouping of patients into disease characterizing subtypes. Subtyping from complex patient data is challenging because of the information heterogeneity and temporal dynamics. Obviously, that the capabilities of the LSTM and GRU neurons can be applied for analyzing longitudinal patient records.

One limitation of the standard LSTM- and GRU-neurons is that they cannot deal with irregular time intervals. But, the time irregularity is common in many healthcare applications. To illustrate this, we can consider patient records, where the time interval between consecutive visits or admissions varies, from days to months and sometimes a year. Such varying time gaps could be indicative of certain impending disease conditions. For example, the frequent admissions might indicate a severe health problem and the records of those visits provide a source to study progression of the condition. If there are months between the two successive records, the dependency on the previous visits or admissions should not play an active role to predict the current outcome.

As was mentioned above, the standard LSTM- and GRU-neurons are designed to handle data with constant elapsed times between consecutive elements of a sequence. Given that time lapse between successive elements in patient records can vary from days to months, the design of the standard LSTM- and GRU-neurons may lead to nonoptimal performance. The novel LSTM-neuron called Time-Aware LSTM-neuron (T-LSTM) to handle irregular time intervals in the longitudinal patient records was proposed in [19]. The subspace decomposition of the cell memory, applied in the T-LSTM-neuron, enables time decay to discount the memory content according to the elapsed time. A herewith, the patient subtyping model uses the proposed T-LSTM-neuron in an auto-encoder to learn a powerful single representation for sequential records of patients, which are then used to cluster patients into clinical subtypes. The architecture of the T-LSTM-neuron captures the underlying structures in the sequences with time irregularities.

The experimental dataset, which can be used to test new methods of maintenance in this sphere, describes so-called the Parkinson's Progression Markers Initiative (PPMI), which is an observational clinical and longitudinal study comprising of evaluations of people with Parkinson's disease (PD) with data for people with high risk, and data for people who are healthy. PPMI aims to identify biomarkers for the progression of Parkinson's disease. PPMI data contains clinical and behavioral assessments, visualization data, and biological samples; therefore, PPMI is a unique PD archive [20].

PPMI is a longitudinal dataset with unstructured elapsed time. In [20], data on 654 patients with idiopathic PD or without PD were studied. The dataset contains the imputed missing values. Also, a unique sign form for categorical values and a data deviation coding of 1 and 0 are used. As a result, the dataset contains 15636 records for 654 patients with an average sequence length of 25 (the minimum sequence length is 3). In [21], data were also classified into features and goals, where the features are associated with the characteristics of the patient, and the goals correspond to the progression of PD.

A herewith, 319 features, such as motor symptoms/complications, cognitive functions, autonomic symptoms, psychotic symptoms, sleep problems, depressive symptoms and anxiety and depression scales in the hospital, and 82 goals, which are associated with motor signs, motor symptoms, cognition and others non-motor factors, were identified [21]. In this dataset, elapsed time is measured in months, and the gap between consecutive patient records is from 1 month to almost 2 years.

It is possible to train and evaluate the regression models, he binary or multi-class classification models for all these problems.

In the future, the first problem will be considered in detail.

3. Approaches to solving the problems of proactive intellectual service

As the analysis of literature shows, in the context of working with TSs, the recurrent neural networks have been actively used. A herewith, various modifications of the recurrent neural networks' models, such as the LSTM network and the GRU network, allows more efficient work with network memory, and, in particular, to solve the so-called problem of the vanishing gradient, which in turn allows them to learn more efficiently (minimizing the problem of overfitting) and solve classification, prediction and analysis problems.

It should be note, that Microsoft provides a template (as part of the Azure Machine Learning offering) which helps data scientists easily build and deploy a predictive maintenance solution [15].

This predictive maintenance template focuses on the techniques used to predict when an in-service machine will fail, so that maintenance can be planned in advance. It allow to solve the following problems [15].

Regression: Predict the Remaining Useful Life (RUL), or Time to Failure (TTF).

Binary classification: Predict if an asset will fail within certain time frame (e.g. days).

Multi-class classification: Predict if an asset will fail in different time windows: e.g., fails in window $[1, w_0]$ days; fails in the window $[w_0+1, w_1]$ days; not fail within w_1 days.

The template includes a collection of pre-configured machine learning modules, as well as custom R scripts in the Execute R Script module, to enable an end-to-end solution from data processing to deploying of the machine learning model. The same solutions can be found in Python.

In this paper we consider various configurations of one- and two-layer networks on the basis of the RNN-, LSTM- and GRU- neurons.

4. Experimental results

The experimental studies were performed for the problem of the aircraft engines' maintenance using the Turbofan Engine Degradation Simulation Data Set [13]. The problem of binary classification of the aircraft engines states was performed. It was necessary to decide: the engine of the considered type is broken or the engine of the considered type is in the working condition during the q-th cycle ($q = 1, 2, ...$).

In order to determine the class of the aircraft engine failure, it is necessary to develop the neural network model and train it at the dataset containing both information about the trouble-free work of the engine and information that a failure has occurred during the work.

For training the neural network model, the *PM* (Predictive Maintenance) dataset was used. It was made publicly available by NASA Ames Research Center [13] and contains: the training set *PM_train* with data on the engine work before the failure; the test set *PM_test* with data on the engine work without registering the failure event; the *GTD* (Ground Truth Data) set with information about the remaining cycles before failure for each engine test set. A herewith, the training and test sets have the similar structure.

The *PM_train* set contains information from several multidimensional time series generated according to the readings of various sensors. It is assumed that each engine starts with varying degrees of the initial wear and production, and this information is unknown to the user. In the training set *PM_train*, it is stated that the engine is working normally at the beginning of each time series. At some point in time during the working cycles, the engine readings begin to deteriorate. Upon reaching the set threshold value, the engine is considered as the unsafe for further work. In other words, each last cycle in each time series can be considered as the failure point of the corresponding engine. One part of this set is used as the training subset, and another part is used as the validation subset during the training of model.

The *PM_test* set also contains information from the several multidimensional time series generated by the readings of various sensors. A herewith, various degrees of wear and deviation at the start of the engine work are also assumed. It is used to estimate the model after training.

At the beginning of each time series, the engine runs normally. At some point in time, engine performance starts to deteriorate, but the last duty cycle for the particular engine is not necessarily a failure point (that is, this set does not show how many cycles this engine can run before it fails).

The neural network model must learn to determine the state when the readings of the engine will be considered as the unsafe for further work.

GTD data determines the number of remaining duty cycles for the engines in the test set.

The development of the neural network model and its training were carried out in Python using the Keras library. This language is widely used for data analysis, and the Keras library is designed specifically for working with the neural networks model. A herewith, the Google Colab was used to minimize the time costs. In general, this allowed to reduce the time costs by 5–6 times.

Previously, on the basis of the training dataset *PM_train*, the training set Train was created by supplementing the training dataset *PM_train* with values for the additional characteristic: the label that defines the class label ("0", if the engine is OK; "1", if the engine is broken). The values of the additional characteristic were determined taking into account the information that the last cycle determines the engine failure point.

During the experiments, the comparative analysis of the different recurrent neural networks configurations was carried out. We considered one-level and two-level configurations (table 1). Figures 1–9 show the graphic dependencies for the model *Accuracies* and the model *Losses* for these configurations.

The structure of the the neural network models in the most cases was defined as follows:

- the *Recurrent layer* with 100 neurons;
- the *Dropout layer* with parameter equal to 0.2 solving the problem of overfitting (it determines the share of the excluded neurons, that is, the neurons that do not contribute to the learning process);
- the *Dense layer* with one neuron and the sigmoidal activation function for solving the classification problem.

If it is necessary, another the *Recurrence layer* and the associated *Dropout layer* are added.

The *Recurrent layer* can be defined using the RNN-, LSTM- or GRU-neurons.

Only one model, the two-layer LSTM+LSTM (100+50 neurons) model, has 50 neurons at the second layer. Figures 10 and 11 show the structure of this model and the form of the output shape with the number of parameters at the each layer respectively [14].

Table 1. Models and their characteristics.

Model structure	Time	Accuracy (Train)	Accuracy (Test)
one-layer RNN (100 neurons)	6 min 36 s	0.995650	0.989247
two-layer RNN+RNN (100+100 neurons)	11 min 10 s	0.995584	0.967742
one-layer GRU(100 neurons)	12 min 43 s	0.994242	0.978495
two-layer GRU+GRU (100+100 neurons)	26 min 15 s	0.996865	0.978495
one-layer LSTM (100 neurons)	19 min 46 s	0.985158	0.978495
two-layer LSTM+LSTM (100+50 neurons)	30 min 30 s	0.993922	0.989247
two-layer LSTM+LSTM (100+100 neurons)	30 min 45 s	0.979720	0.967742
two-layer LSTM+GRU (100+100 neurons)	28 min 56 s	0.9968652	0.978495
two-layer GRU+LSTM (100+100 neurons)	28 min 43 s	0.995970	0.978495

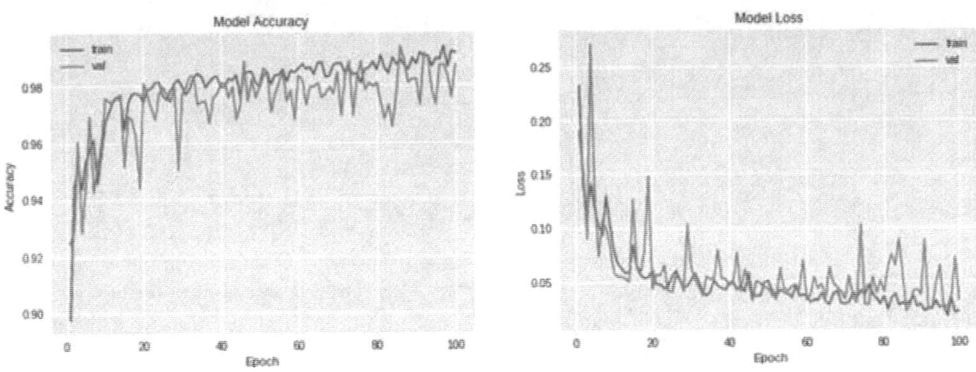

Figure 1. The graphic dependencies for the one-layer RNN model.

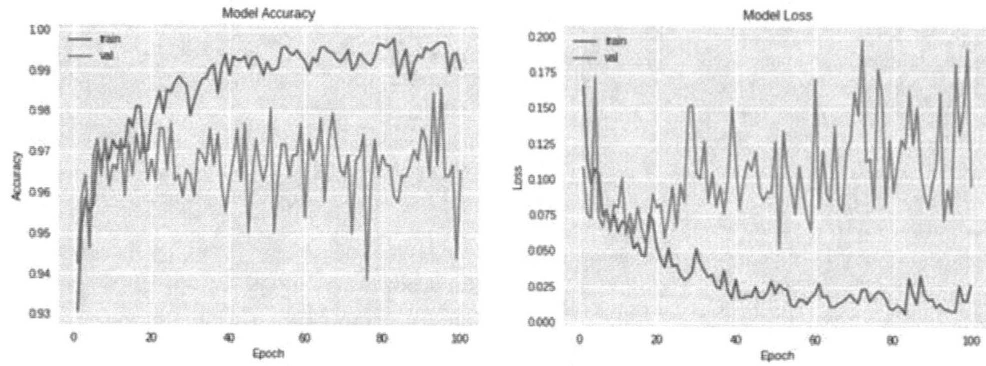

Figure 2. The graphic dependencies for the two-layer RNN model.

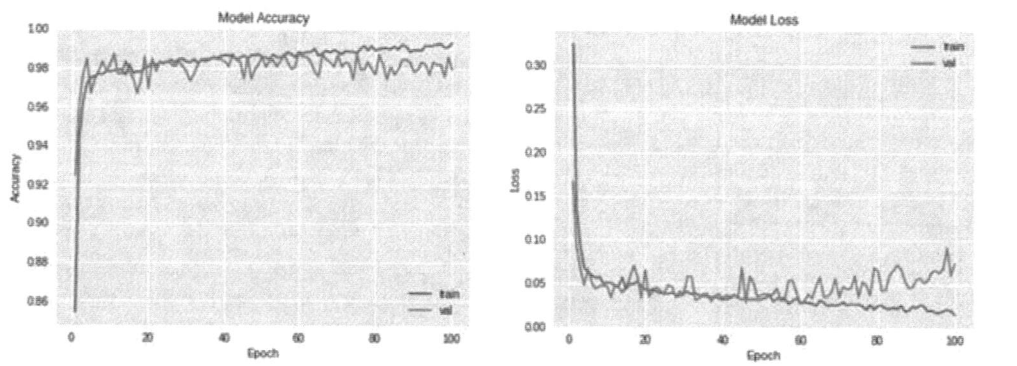

Figure 3. The graphic dependencies for the one-layer GRU model.

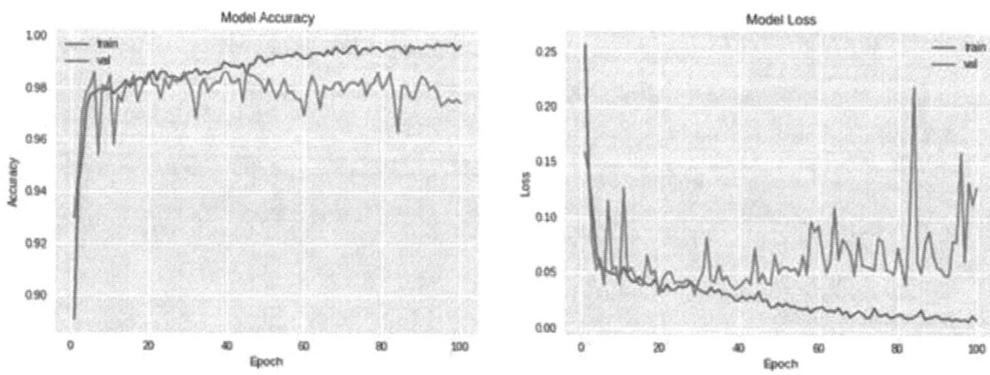

Figure 4. The graphic dependencies for the two-layer GRU model.

During the experiments, we used the *Adam* algorithm as the optimization algorithm [22]. Also, we considered the *Accuracy indicator* as the objective function, and the *binary cross-entropy function* which returns the classification error as the loss logistic function *Loss*. We used 100 epochs to observe the model development process.

Table 1 shows the model structures, time costs, and values of the *Accuracies indicator* at the training and test (validate) sets.

It is known, that the results of training the neural network substantially depend on the correct initialization of the neuron weights. A herewith, it is required that the found solution will always be close to suboptimal.

The minimum development time, as was expected, corresponds to one-layer and two-layer RNN network models (figure 1 and 2). A herewith, it is very difficult to track overfitting in the case of one-layer RNN network (figure 1) when overfitting occurs and say when it is advisable to perform the early stopping of the learning process. In this context, one-layer GRU (figure 3) and LSTM (figure 5) network models behave noticeably better. The two-layer RNN network model (figure 2) also behaves better in this context than the one-layer RNN network model (figure 1), but it has one of the worst value of the *Accuracy* indicator at the test set.

The two-layer GRU network model (figure 4) allows making the stopping earlier than the one-layer GRU network model (figure 3) halving the number of epochs (about 30 epochs against 60).

The LSTM network models require fewer epochs for the early stopping: about 40 epochs for the one-layer LSTM network model (figure 5) and about 30 epochs for the two-layer LSTM network model (figure 6 and 7). A herewith, the time costs increase slightly. Selection of the optimal number of neurons in the second layer (figures 6 and 7) requires additional attention: it is important to minimize the number of neurons (and, therefore, the time costs on developing the model) without impairing the *Accuracy indicator* of the model.

The hybrid network models (figures 8 and 9) work similarly in the context of accuracy, but, at first glance, the two-layer LSTM+GRU network model (figure 9) allows the earlier stop, however ,the additional research is needed in this direction.

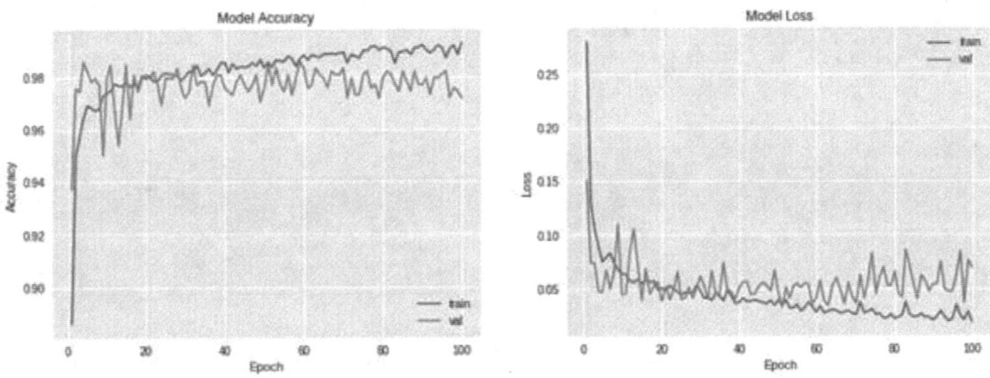

Figure 5. The graphic dependencies for the one-layer LSTM model.

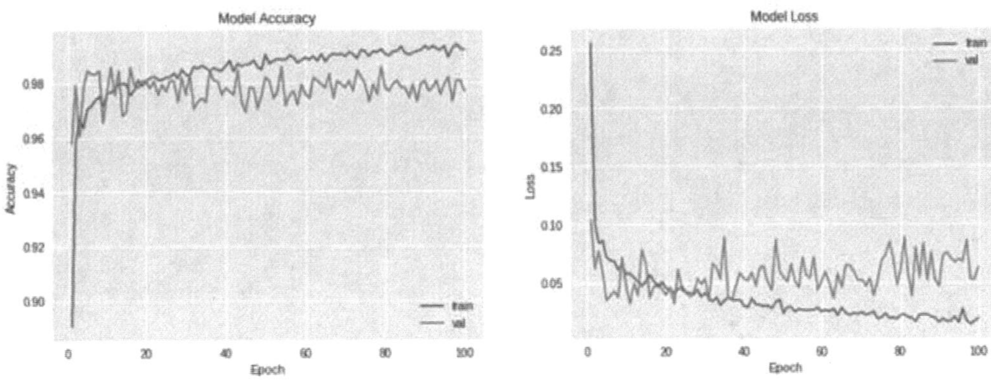

Figure 6. The graphic dependencies for the two-layer LSTM (100+50) model.

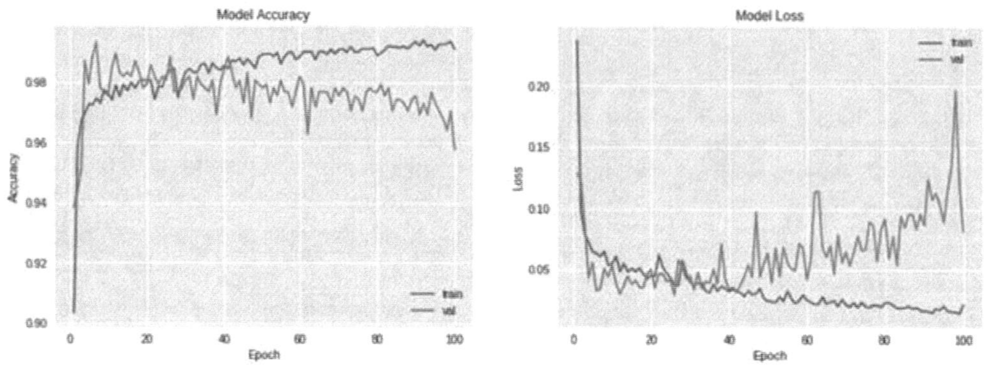

Figure 7. The graphic dependencies for the two-layer LSTM (100+100 neurons) model.

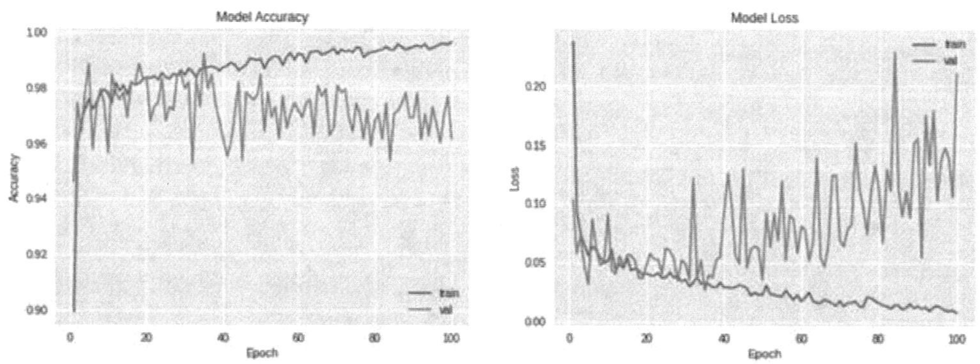

Figure 8. The graphic dependencies for the two-layer GRU+LSTM model.

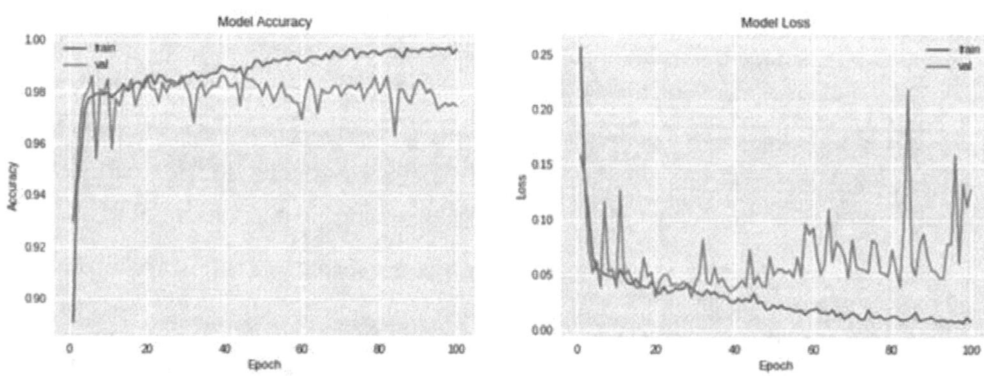

Figure 9. The graphic dependencies for the two-layer LSTM+GRU model.

A comparative analysis of the results of training of the neural networks models shows the equal or close values of the *Accuracy indicator* for the GRU- and LSTM- network models. A herewith, the learning speed of the network model based on the GRU-neurons exceeds the learning speed of the network model based on the LSTM-neurons (as for one-layer network model as for two-layer network model).

```
# build the network
nb_features = seq_array.shape[2]
nb_out = label_array.shape[1]

model = Sequential()

model.add(LSTM(
         input_shape=(sequence_length, nb_features),
         units=100,
         return_sequences=True))
model.add(Dropout(0.2))

model.add(LSTM(
          units=50,
          return_sequences=False))
model.add(Dropout(0.2))

model.add(Dense(units=nb_out, activation='sigmoid'))
model.compile(loss='binary_crossentropy', optimizer='adam', metrics=['accuracy'])
```

Figure 10. The form of the output shape with the number of parameters at the each layer of the two-layer LSTM+LSTM (100+50 neurons) model.

Layer (type)	Output Shape	Param #
lstm_1 (LSTM)	(None, 50, 100)	50400
dropout_1 (Dropout)	(None, 50, 100)	0
lstm_2 (LSTM)	(None, 50)	30200
dropout_2 (Dropout)	(None, 50)	0
dense_1 (Dense)	(None, 1)	51

```
Total params: 80,651
Trainable params: 80,651
Non-trainable params: 0

None
```

Figure 11. The structure of the two-layer LSTM+LSTM (100+50 neurons) modelMoreover, these neural networks solve the problem of overfitting and early stopping during the development of the neural network.

5. Conclusion

The experimental results show, that the LSTM- and GRU-networks demonstrate their high efficiency in solving the classification problem of the probable class of failure in the work of aircraft engines. It is planned to explore the possibilities of these networks to solve the classification problem of the probable class of failure of equipment taking into account the aspects of multicriteria optimization.

References

[1] Vapnik V 1998 *Statistical Learning Theory*

[2] Demidova L A, Nikulchev E V and Sokolova Y S 2015 *International Review on Modelling and Simulations* **4** 446–57

[3] Demidova L, Nikulchev E and Sokolova Y 2016 *International Journal of Advanced Computer Science and Applications (IJACSA)* **7** 16–24

[4] Scheibelhofer P, Leditzky G, Berger M, Schröttner P, Gleispach D and G. Hayderer 2011 *Statistical Methods Applied in Microelectronics*

[5] Elghazel Wiem, Medjaher Kamal, Zerhouni Noureddine, Bahi Jacques, Farhat Ahmad, et al. 2015 *IEEE Aerospace Conference* pp 1–9

[6] Hopfield J and Tank D 1985 *Biological Cybernetics* **52** 141–52

[7] Rosenblatt F 1992 *Principle of neurodynamics*

[8] Williams R and Zipser D 1989 *Neural Computers* **1** 270–80

[9] Sigov A S, Andrianova E G, Zhukov D O, Zykov S V and Tarasov I E 2019 *Rossiyskiy tekhnologicheskiy zhurnal (Russian Technological Journal)* **1** 5–37 DOI: 10.32362/2500-316X-2019-7-1-5-37

[10] Ding H, Yang L and Yang A Zh 2019 *IEEE Access* **7** 10.1109/ACCESS.2019.2933676

[11] Hochreiter S and Schmidhuber 1997 *Neural computation* **8** 1735–80

[12] Aydın O and Guldamlasioglu S 2017 4th International Conference on Electrical and Electronic Engineering (ICEEE) doi: 10.1109/ICEEE2.2017.7935834

[13] Saxena A and Goebel K *NASA Ames Prognostics Data Repository* NASA Ames Research Center, Moffett Field, CA https://ti.arc.nasa.gov/tech/dash/pcoe/prognostic-data-repository/

[14] https://github.com/Azure/lstms_for_predictive_maintenance

[15] https://gallery.azure.ai/Collection/Predictive-Maintenance-Template-3

[16] Lima F D S, Amaral G M R, Leite L G M, Gomes J P P and Machado J C 2017 *Brazilian Conference on Intelligent Systems* pp 222–7

[17] Anantharaman P, Qiao M and Jadav D 2018 *IEEE International Congress on Big Data (BigData Congress)* pp 251–4

[18] https://www.backblaze.com/b2/hard-drive-test-data.html

[19] Baytas I M, Xiao C, Zhang X, Wang F, Jain A K and Zhou J 2017 *Proceedings of the 23rd ACM SIGKDD international conference on knowledge discovery and data mining* pp 65–74

[20] Che Ch, Xiao C, Liang J, Jin B, Zhou J and Wang F 2017 *Proceedings of the 2017 SIAM International Conference on Data Mining* pp 198–206

[21] Dinov I D, Heavner B, Tang M, Glusman G, Chard K and Darcy M et al. *2016 PLoS ONE* https://doi.org/10.1371/journal.pone.0157077

[22] Kingma D P and Ba J L 2014 Preprint/arXiv.org № 1412.6980

Special Aspects of Quality Assurance in the Design, Manufacture, Testing of Aerospace Engineering Products

A Ya Dmitriev[1], R S Zagidulin[1] and T A Mitroshkina[1]

[1]Samara National Research University named after academician S.P. Korolev (Samara University), Samara, Russia

E-mail: dmitriev57@rambler.ru

Received xxxxxx
Accepted for publication xxxxxx
Published xxxxxx

Abstract

A quality assurance model is proposed for the design, manufacture, testing of aerospace products based on robust design approaches and the widespread use of additive technologies. Examples of developing proposals and optimizing the design of aerospace engineering products based on the presented are given. Tests of optimized prototypes of products made by selective laser sintering show positive results.

Keywords: quality, quality assurance, design, robust design, FMEA, QFD, engineering, additive manufacturing, model

1. Introduction

Aerospace engineering companies use quality management approaches based on quality management standards (ISO 9001, AS / EN 9100) and proven engineering methods for quality management, such as APQP advanced product planning process, SPC statistical process control, measurement system analysis MSA, QFD quality function deployment, analysis of the types, causes and consequences of potential FMEA nonconformities. The application of these methods, along with robust approaches to design, three-dimensional design systems and engineering analysis (for example, ANSIS), makes it possible not only to produce products that meet the requirements, but also significantly reduce costs at the development and design stages [1, 2].

At the same time, the potential of traditional Russian approaches and technologies for the design and manufacture of aerospace engineering products has been almost exhausted. In recent years, the introduction of additive production of aerospace technology has accelerated, especially thanks to advances in the production of metal powders for 3D printing [3]. Additive technologies present unique opportunities for the design and manufacture of promising aerospace technology products. Such well-known manufacturing companies as Boeing, Airbus, SpaceX, Blue Origin, Rocket Lab and others, every year in the designs of their products increase the number of parts made using 3D printing [3–5].

2. Examples of the implementation of modern approaches to quality assurance in the aerospace industry

Integration into a single system of robust approaches, engineering methods, software systems for three-dimensional design and finite element analysis, 3D printing makes it possible to create and produce promising products of a completely new concept.

2.1 Designing special technological equipment for implementing tests simulating the effects of aerodynamic loads of the assembly and protective unit of the scientific and energy module

In [6], a detailed description is given of the application of the QFD and FMEA methods in the design of a special technological tooling by a Russian aerospace industry company to implement tests simulating the effects of aerodynamic loads of an assembly-protective unit of a scientific and energy module. Using the QFD method [1, 2], the characteristics that most determine the quality of the tooling were revealed. As a result of QFD of the first level, the most important design characteristics of special technological equipment have been identified that have the greatest impact on the strength and rigidity of the structure, as well as the fulfillment of other technical requirements: tool material, thickness of ribs and sheathing sheets. During the following QFD levels, the most important design characteristics and production process parameters were identified. To ensure the quality of the design, an additional QFD analysis was carried out, which examined the parameters that ensure the quality of the design of technological equipment, as a result of which the key parameters of the design process of special technological equipment were identified: the need for calculations in the ANSYS system, the use of 3D-model geometry in the tasks of technological preparation of production.

An FMEA analysis [1] was carried out for the design of special technological equipment, as a result of which actions to prevent and reduce risks were identified. Potential design inconsistencies "Necessary structural strength", "Material stiffness and the design of elastic pressure bags distort pressure" and others, which can lead not only to incorrect test results, but to damage to the tooling, are examined.

Taking into account the requirements laid down in the terms of reference for the design of special technological equipment for testing, and the results of the analysis by QFD and FMEA methods, technological equipment was designed. The calculations for the strength and rigidity of the tooling structure were made in the ANSYS finite element analysis software system. The estimated safety factor is not lower than required, therefore, the design meets the specified reliability requirements in real operating conditions.

2.2. Improving the performance characteristics of light and ultra-light launch vehicles (LV) with the introduction of modern approaches to quality assurance and additive technologies

Nowadays Russian Federation works at improving tactical and technical characteristics of light and ultra-light launch vehicles (LV) with the introduction of modern approaches to ensuring quality and additive technologies. The work is aimed at increasing the competitiveness of launch vehicles in the launch services market. The development of LVs of the light class and ultralight class is due to the currently observed tendency to increase the need to launch small spacecraft.

Competitors of this project to one degree or another are the projects of light and ultra-light carriers of the USA, China, Japan (Firefly Space Systems, Rocket Lab, etc.). The reduction in the mass and dimensions of the spacecraft and the spacecraft is a focused technical policy aimed at lowering the cost and increasing the availability of space services for the mass consumer. Accordingly, additive technologies are already used in a number of projects [5].

The analysis was carried out using the QFD quality function deployment method at the stage of developing technical proposals. At this stage, all the technical characteristics (mass, geometric, energy, etc.) of the future rocket are described. As a result of QFD, the most important implementation methods were identified for improving the tactical and technical characteristics of the launch vehicle, which have the greatest impact on the reliability and cost of putting the payload into orbit, as well as other requirements: reducing the launch vehicle mass by optimizing the design and choosing the launch vehicle manufacturing technology.

Decrease in the mass of the launch vehicle due to structural optimization and the choice of manufacturing technology for the launch vehicle require topological optimization of the design using such engineering analysis systems as ANSYS Topology Optimization, Siemens Solid Edge, Function Driven Generative Designer, and the use of advanced technologies for the manufacture of parts and assemblies, such as 3D printing.

Further, to assess risks and select areas for improvement, an FMEA analysis was carried out for the construction of a light class LV. Based on the results of the FMEA, proposals were developed to reduce risks, including: topological optimization of the design of highly loaded parts and assemblies, 3D printing of parts and assemblies obtained as a result of topological optimization, field testing (static, dynamic, thermal vacuum) of parts and assemblies manufactured on 3D printer before entering them into the design of the LV.

New features of three-dimensional design systems, topological design optimization and the use of additive technologies make it possible to obtain ultralight parts and assemblies (weight reduction in some cases reaches 55%), not inferior in strength to parts and assemblies that are currently in operation as part of the LV design. In addition, topological optimization and additive technologies can reduce the number of parts in the assembly units of the product design, the time and cost of technological preparation of production.

Before the introduction of additive technologies in the production of aerospace engineering, topological optimization, 3D printing and full-scale testing of prototypes

IOP Conf. Series: Materials Science and Engineering **714** (2020) 012006 doi:10.1088/1757-899X/714/1/012006

(static, dynamic, thermal vacuum) are carried out. Next, the technological process of 3D printing is developed for each group of similar parts and assemblies taking into account control and destabilizing factors, followed by the development of technological instructions and an organization standard.

At the stage of introducing additive technologies into the production of aerospace equipment, the "Fork" part was used as a prototype. Optimization was performed (in the ANSYS system) and the part was manufactured by selective laser sintering. The material of the grown part is 316L steel (chromium-nickel steel). In fig. Figure 3 shows the initial geometry of the "Fork" part and a variant of optimized geometry. The weight reduction of the optimized part was 40%.

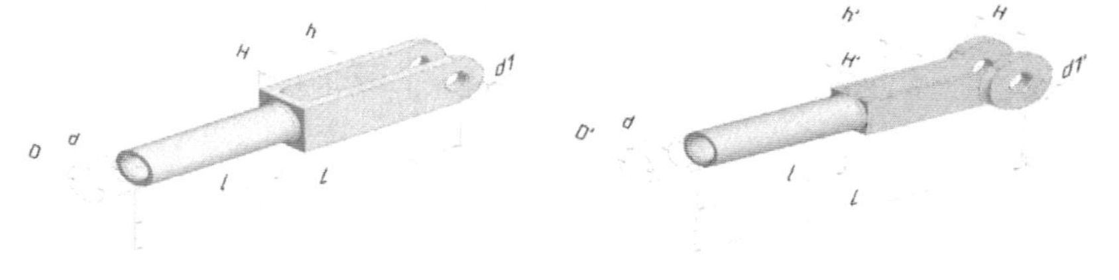

Figure 1. The geometry of the "Fork" part: (a) source; (b) optimized.

Figure 2. "Fork" part on the substrate of the 3D printer.

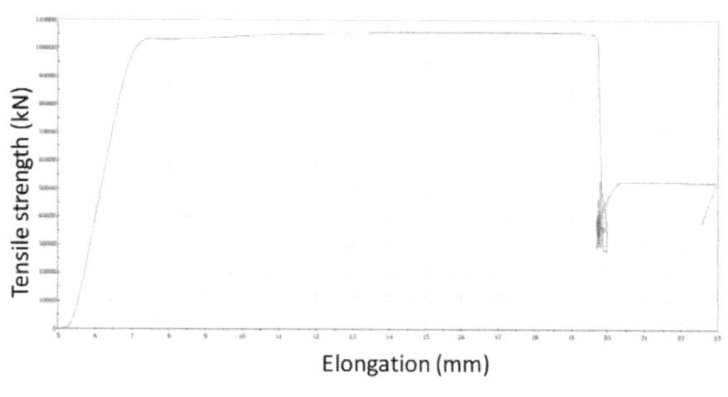

Figure 3. "Fork" part Tensile Test Results.

Figure 2 shows a printed part on a 3D printer substrate. The "Fork" part made on a 3D printer has successfully passed thermal vacuum and static tests (Figure 3), ultimate tensile strength is more 100kN.

The combination of software systems for design and optimization and the use of additive technologies allow us to implement a new approach, different from the traditional approach, to the design and manufacture of products — bionic design. Figure 4 shows a bracket for installing sensitive elements of the spacecraft for remote sensing of the Earth.

Bionic design is the design and subsequent manufacture of products with an extremely complex design geometry, similar to the structure found in living nature.

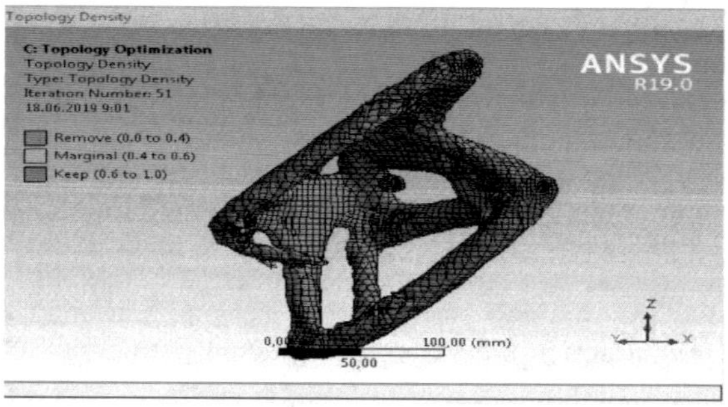

Figure 4. Bracket for installing sensitive elements.

3. The modern model of quality assurance in the design, manufacture and testing of aerospace engineering

The modern model of quality assurance in the design, manufacture and testing of aerospace engineering products includes the use of quality assurance methods, software systems for design and optimization, and the use of additive technologies (Figure 5).

The presented model is based on a new approach to the design and manufacture of products — bionic design and on approaches to long-term planning of product quality adopted in the aerospace industry. The model is aimed at reducing costs and non-production losses both at the early stages of design and at the stages of production and testing, based on modern QFD and FMEA quality assurance methods and additive production approaches.

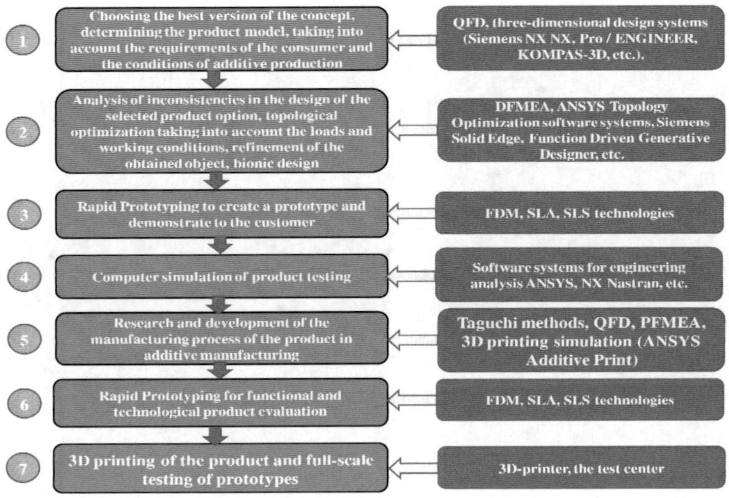

Figure 5. The modern model of quality assurance in the design, manufacture and testing of aerospace engineering.

4. Conclusions and discussion

The aerospace industry is a promising industry worldwide. Aerospace technology manufacturers apply modern approaches to ensuring quality throughout the entire product life cycle. The key areas of implementation and development of the presented model of quality assurance in the design and manufacture of aerospace technology are robust approaches and parametric identification of the product model [1, 2, 7, 8]. The main direction of further use of the expected results is the testing of the proposed model for the development of new products of aerospace technology.

References

[1] Dmitriev A and Mitroshkina T 2016 *ITM Web of Conferences* **6** 02005 doi: 10.1051/itmconf/20160602005

[2] Dmitriev A, Mitroshkina T and Rogachev G 2017 *ITM Web of Conferences* **10** 01001 doi: 10.1051/itmconf/20171001001

[3] Library OECD. How Space Contributes to the Global Economy. Chapter 4. Digital (r)evolution in manufacturing and in the production of space systems https://www.oecd-ilibrary.org/sites/bd1505f8-en/index.html?itemId=/content/component/bd1505f8-en&mimeType=text/html

[4] Russell K 2017 Thales Alenia Space saves time, money with 3D printing *Via Satellite* 27 June https://www.satellitetoday.com/innovation/2017/06/27/3d-printing-future-satellite-manufacturing/

[5] Klyushnikov V Y 2019 *Aerospace Sphere Journal* **3** 58–71 doi: 10.30981/2587-7992-2019-100-3-58-71

[6] Zigadullin R S, Barinov P V, Burkova V A, Glushkov S V and Mitroshkina T A 2019 *Quality and life* **2** 45–53 doi: 10.34214/2312-5209-2019-22-2-44-53

[7] Dmitriev A and Mitroshkina T 2019 Improving the efficiency of aviation products design based on international standards and robust approaches *IOP Conference Series: Materials Science and Engineering* **476** 1 DOI: https://doi.org/10.1088/1757-899X/476/1/012009

[8] Dmitriev A Ya and Mitroshkina T A 2015 *Ontology of designing* **3** 313–27 doi: 10.18287/2223-9537-2015-5-3-313-327

Buckling Problem Statement and Approaches to Buckling Problem Investigation of Structurally-Anisotropic Aircraft Panels Made from Composite Materials

L M Gavva[1], V V Firsanov[1] and A N Korochkov[1]

[1] Moscow Aviation Institute, Moscow, Russia

E-mail: rva101@mail.ru

Received xxxxxx
Accepted for publication xxxxxx
Published xxxxxx

Abstract

Aircraft composite structure design in the field of production technology is the outlook research trend. New mathematical model relations for the buckling investigation of structurally-anisotropic panels comprising composite materials are presented. The primary scientific novelty is the further development of the theory of thin-walled elastic ribs related to the contact problem for the skin and the rib with an improved rib model. One considers the residual thermal stresses and the preliminary tension of the reinforcing fibers with respect to panel production technology. The buckling problem results in the boundary value problem when solving for the eighth order partial derivative equation in the rectangular field. The solution is designed by a double trigonometric series and by a unitary trigonometric series. The results of testing series are presented.

Keywords: panels made of composite materials, eccentric longitudinal and lateral set, force and technology temperature action, pre-critical stress state, buckling, bending mode, torsion mode

1. Introduction

The buckling problems of a flat rectangular multilayer panel made from polymer fiber composite materials with the eccentric longitudinal and lateral stiffening set are considered. The panels are subjected to the distributed constant compressive loading applied to the edges in the casing plane in the stationary temperature field. The boundary conditions at the two opposite sides of the contour are assumed to be the particular case with conformable boundary restrictions for the plane problem and problem of bending, the solution in closed form is designed by a trigonometric series.

One should take into consideration the technological factors occurring in the fabrication of composites, namely, residual thermal stresses arising during cooling after hardening and pre-stressed tension of reinforcing fibers that is performed in order to increase the bearing strength of the structure [1].

The refined statement of the buckling problems has been formulated subject to the pre-critical stress state in the compression of flat rectangular multiplied panels made of polymer fiber composite materials.

The papers [2–38] describe buckling problems of structurally-anisotropic composite panels: buckling problem statement, buckling problem statement with thermal and force loading, buckling problem statement subject to production technology, analytical methods to solve buckling problems, numerical methods to solve buckling problems, test investigations. The presented mathematical model is seemed to be new one.

The critical force calculations for the general bending mode of the thin-walled system buckling and the critical force calculations for the multi-wave torsion buckling are of the most actual interest in accordance with traditional design practices. In both cases, bending is integral with the plane stress state. The schematization of the panel as structurally-anisotropic has been proposed as a design model when the stress-strain state and critical forces of total bending mode of buckling are determined. For a multi-wave torsion buckling study, one should use the generalized function techniques.

The scientific novelty of this research reflects the further development of the theory of thin-walled elastic ribs related to the contact problem for the skin and rib with an improved rib torsion model. The aim of this study is the buckling problem statement and the approaches to solve this problem in view of the non-uniform pre-critical stress state and production technology. The buckling problem results in the boundary value problem when solving for the eighth order partial derivative equation in the rectangular field. The buckling problem statement and the proposed approaches to solve this problem are new and are of the interest from the design and manufacture of aircraft outlook specimen made from modern composite materials. The aim of this study is to confirm the authenticity of the presented mathematical model and to verify the research results.

2. Buckling of structurally-anisotropic composite panels. Problem Statement

Both cases, the buckling problem as the problem of the pre-critical stress state cannot be divided into a plane part and a plate bending.

The eighth order differential equation is resolved for the buckling problems (e.g. Firsanov and Gavva 2017). This equation is designed as an equilibrium differential equation with the effect of the reduced loading connected with the normal forces N_x, N_y and tangential forces N_{xy}, N_{yx}.

The deflection function *w(x,y)* is coupled with the potential function *Φ(x,y)*.

For the composite panel of orthotropic structure the left-hand part of the eighth order differential equation is limited by the even-numbered partial derivatives of *Φ(x,y)* but the odd-numbered partial derivatives in the right-hand part are connected with the shear

$$\frac{K_{80}}{a^8}\frac{\partial^8\Phi}{\partial x^8}+\frac{K_{62}}{a^6b^2}\frac{\partial^8\Phi}{\partial x^6\partial y^2}+\frac{K_{44}}{a^4b^4}\frac{\partial^8\Phi}{\partial x^4\partial y^4}+\frac{K_{26}}{a^2b^6}\frac{\partial^8\Phi}{\partial x^2\partial y^6}+\frac{K_{08}}{b^8}\frac{\partial^8\Phi}{\partial y^8}=$$
$$=\begin{bmatrix} \frac{N_xR_{40}}{a^6}\frac{\partial^6\Phi}{\partial x^6} & +\frac{(N_{xy}+N_{yx})R_{40}}{a^5b}\frac{\partial^6\Phi}{\partial x^5\partial y}+ \\ +\frac{(N_xR_{22}+N_yR_{40})}{a^4b^2}\frac{\partial^6\Phi}{\partial x^4\partial y^2} & +\frac{(N_{xy}+N_{yx})R_{22}}{a^3b^3}\frac{\partial^6\Phi}{\partial x^3\partial y^3}+ \\ +\frac{(N_xR_{04}+N_yR_{22})}{a^2b^4}\frac{\partial^6\Phi}{\partial x^2\partial y^4} & +\frac{(N_{xy}+N_{yx})R_{04}}{ab^5}\frac{\partial^6\Phi}{\partial x\partial y^5}+ \\ +\frac{N_yR_{04}}{b^6}\frac{\partial^6\Phi}{\partial y^6} & \end{bmatrix} \quad (1)$$

$x=x/a$, $y=y/b$ are the dimensionless coordinates related to the panel length *a* and to its width *b*.

The coefficients R_{ij}, $i=4,3,\ ...,\ 0;\ j=0,1,\ ...,\ 4$ in the relation formulas and the coefficients K_{ij}, $i=8,7,\ ...,\ 0;\ j=0,1,\ ...,\ 8$ in the resolving equation (1) are the constant values, which depend on the elastic characteristics of the material and geometrical structure parameters.

All components of the stress-strain state including the inner force factors are related with the potential function *Φ(x, y)* as

$$\left.\begin{aligned} N_x=L_{Nx}\Phi-N_x^T-N_x^H,\quad N_y=L_{Ny}\Phi-N_y^T-N_y^H \\ N_{xy}=L_{Nxy}\Phi-N_{xy}^T-N_{xy}^H,\quad N_{yx}=L_{Nyx}\Phi-N_{yx}^T-N_{yx}^H \end{aligned}\right\} \quad (2)$$

The linear differential operator for the orthotropic structure, for example

$$L_{Nx}=P_{60}^x\frac{\partial^6}{\partial x^6}+P_{42}^x\frac{\partial^6}{\partial x^4\partial y^2}+P_{24}^x\frac{\partial^6}{\partial x^2\partial y^4}+P_{06}^x\frac{\partial^6}{\partial y^6},$$

The thermal forces and moments are $N_x^T, N_y^T, N_{xy}^T, N_{yx}^T$, , the tension forces and moments of the composite fibers are $N_x^H, N_y^H, N_{xy}^H, N_{yx}^H$. The coefficients P_{ij}^x, $i=6,4,2,0;$ $j=0,2,4,6$ as the coefficients in the relation formulas, depend on the elastic characteristics of the material and geometrical structure parameters.

The buckling problem for a structurally-anisotropic composite panel is a nonlinear one according to equation (1) and connection formulas (2). The linearization method is used to determine the critical forces. First, one considers the stress-strain state of the structure at compression, namely, the pre-critical main stress state being complicated as it is not divided into the plane problem and bending problem according to the mathematical model proposed. It is necessary to determine the distribution law of the normal and shear inner forces caused by the external loading. Then one considers the buckling problem as the proper value problem to obtain the additional displacement of the basis surface.

IOP Conf. Series: Materials Science and Engineering **714** (2020) 012007 doi:10.1088/1757-899X/714/1/012007

3. Pre-critical stress-strain state of structurally-anisotropic composite panels in compression

We consider the pre-buckling stress state of a flat rectangular composite panel with the eccentric stiffening set being orthotropic one. The panel is subjected to the uniform distributed normal compressive loading with P intensity applied to the lateral opposite sides in the skin plane. Boundary restrictions satisfy the hinging condition in respect to bending and the sliding constraint condition in the tangential direction for the plane problem when the panel is loaded by the shear force flows along its longitudinal edges.

The normal forces N_x, corresponding to the pre-critical stress state of the stiffened composite panel compressed along the x axis are distributed as

$$N_x = P \sum_{i=1,3,5...}^{\infty} \left\{ \left[\sum_{L=1}^{4} (N_x)_{iL} \, ch(\lambda_{iL} x) \right] - \frac{(N_x^T)_i}{P} - \frac{(N_x^H)_i}{P} \right\} \sin(i\pi y) \quad (3)$$

$(N_x)_{iL}$ are the coefficients of single trigonometric series for the normal forces N_x, known after the determination of the constants of single trigonometric series, $(N_x^T)_i$, $(N_x^H)_i$ are the coefficients of single trigonometric series for the thermal and tension forces.

Here $\lambda_{iL} = z_L \lambda_{iy} a$, $\lambda_{iy} = \frac{i\pi}{b}$, z_L are the roots of the corresponding characteristic polynomial and are calculated using the MATLAB operating environment.

The variations of the transverse forces N_y and shear forces N_{xy}, N_{yx} are obtained analog to (3).

One can estimate the influence of production technology factors on the bearing strength of structurally-anisotropic composite panels if the non-uniform pre-critical stressed state is considered, boundary conditions are non-conformable, and the solution is formed by a unitary trigonometric series.

4. Buckling of structurally-anisotropic composite panels subject to pre-critical stress state

The linear buckling problem is formulated as the proper value problem.

The general differential equation of the curved surface (1) determines the additional equilibrium state of the structure in view of the initial pre-buckling state.

The integral of this equation satisfying to the uniform boundary conditions may be approximated by a double trigonometric series. But it is not feasible to solve the problem in close form as the method is reduced with an infinite system of the linear algebraic equations. The one-member approximation is considered as the first solution.

The critical force formula is designed with the orthogonalization procedure of the general differential equation of the curved surface. The expression

$$P = P^* \left\{ \frac{\pi}{\sum_{i=1,3,5...}^{\infty} \left(-\frac{2}{i} - \frac{1}{2n-i} + \frac{1}{2n+i} \right) \left\{ \left[\sum_{L=1}^{4} (N_x)_{iL} \frac{\left[(m\pi)^2 - \lambda_{iL}^2 \right]}{\left[(m\pi)^2 + \lambda_{iL}^2 \right]} \frac{sh\lambda_{iL}}{\lambda_{iL}} \right] - \frac{(N_x^T)_i}{P} - \frac{(N_x^H)_i}{P} \right\}} \right\} \quad (4)$$

provides the range of values P for additional possible deformation of the base surface at m = 1, 2, 3,…, n = 1, 2, 3,…, m and n are the wave parameters.

The P with «*» is the critical force calculated with the main uniform pre-buckling stress state (e.g. Firsanov and Gavva 2017)

$$P^* = \frac{\pi^2}{b^2} \frac{K_{80}\left(\frac{m}{c}\right)^8 + K_{62}\left(\frac{m}{c}\right)^6 n^2 + K_{44}\left(\frac{m}{c}\right)^4 n^4 + K_{26}\left(\frac{m}{c}\right)^2 n^6 + K_{08} n^8}{\left[R_{40}\left(\frac{m}{c}\right)^4 + R_{22}\left(\frac{m}{c}\right)^2 n^2 + R_{04} n^4 \right] \left(\frac{m}{c}\right)^2},$$

for the general bending mode of buckling, the panel side ratio $c = 2a/b$, a, b are the panel half-length and width, correspondingly.

The formula for the critical loading P of the multi-wave torsion buckling problem coincides to this formula within the coefficients K_{ij} and R_{ij}. The coefficients $\hat{K}_{ij}$, $i, j = 0, 2, 4, 6, 8$ and $\hat{R}_{ij}$, $i, j = 0, 2, 4$, are determined by generalized stiffness characteristics while stiffness averaging for the elements of longitudinal set up to the skin is replaced with the discrete characteristics

$$\frac{1}{c_1} \to \begin{cases} \frac{2}{b} \sum_{i=1}^{N} \sin^2(n\pi y_i) \\ \frac{2}{b} \sum_{i=1}^{N} \cos^2(n\pi y_i) \end{cases} \to \hat{K}_{ij}, \hat{R}_{ij},$$

c_1 is the stringer distance; y_i is the coordinate y of the discrete stringer.

It is allowable to estimate the influence of production technology factors on the bearing strength of the structurally-anisotropic composite panels if the complicated pre-critical stress state is considered and the solution is derived in a single approximation of the trigonometric series. One considers the residual thermal stresses arising during cooling after hardening and the pre-stressed tension of the reinforcing fibers with respect to panel production technology.

The step by step method is used to determine the critical forces. The critical force P with «*» calculated with the uniform pre-buckling stress state is proposed as an initial first approach.

5. Buckling of structurally-anisotropic composite panels subject to uniform pre-critical stress state. General boundary problem

One considered the general boundary problem at two opposite sides for buckling of structurally-anisotropic composite panels.

The integral of the equation (1) satisfying to the boundary conditions of general type along lateral edges may be approximated by a single trigonometric series.

This case the step by step method is used to determine the critical forces according to Figure 1.

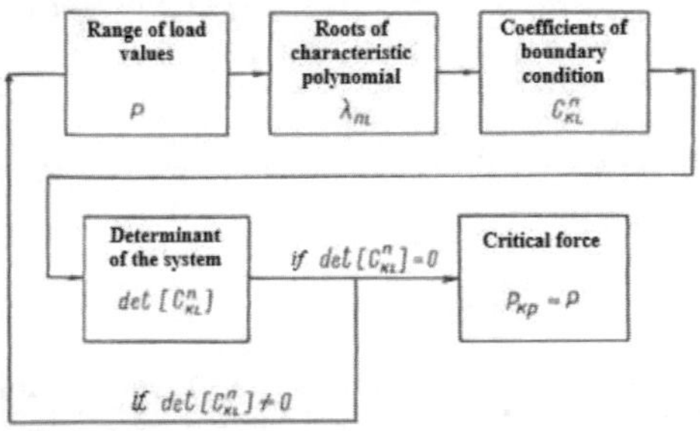

Figure 1. Step by step method for critical forces. Panels compressed in the longitudinal direction. General boundary problem at two opposite sides.

6. Buckling of structurally-anisotropic composite panels subject to pre-critical stress state. Results and Discussion

A computer program package is developed using the MATLAB operating environment. The computer program package has been utilized for the determination of the critical forces subject to the uniform and non-uniform pre-buckling stress state and for multi-criteria optimization of the design of structurally-anisotropic aircraft composite panels.

The results of determining the critical parameters for rectangular carbon plastic panels eccentrically stiffened and compressed in the longitudinal direction are presented as an example in view of the uniform pre-buckling stress state.

The testing series of uniform compressed stiffened composite panels for carrying the objects to the moment of stability loss (Figure 2) have been made using the special fixture for the proposed mathematical model verification.

The refined theoretical results and the experimental data are in agreement qualitatively with respect to buckling modes and quantitatively with respect to critical stresses within 12–13% if pre-buckling stress state is considered uniform. And they reveal a higher precision of 8–10% in view of non-uniform main pre-buckling stress state. Thus, it confirms the authenticity of the presented mathematical model.

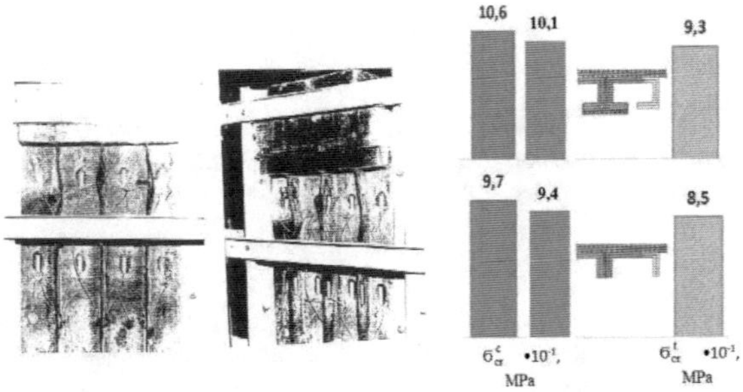

(a) (b)

Figure 2. Stringer panels compressed in the longitudinal direction: (a) experimental specimen; (b) comparison of the test and calculation results for critical stresses.

The testing series of uniform compressed stiffened panels comprising composite bundles for carrying the objects to the moment of stability loss (Figure 3) have been made using the special fixture for the proposed mathematical model verification.

Figure 3. Stringer panels compressed in the longitudinal direction.

The refined theoretical results and the experimental data are in agreement qualitatively with respect to buckling modes and quantitatively with respect to critical forces within 19 - 20% if pre-buckling stress state is considered uniform. The stringer cross section of the panel specimen is non-symmetric, but the numerical results were obtained for the panels with the symmetric cross section of the stringer. Thus, it also confirms the authenticity of the presented mathematical model.

7. Conclusion

Since the solution is obtained by analytical methods, the calculation time is minimal. This is of interest from the perspective of practical design using parametric analysis. The results of the stress analysis calculations, as well as the results of the buckling analysis calculations, offer opportunities for reducing and optimizing the weight characteristics of aircraft elements.

Acknowledgements

The study was performed in the framework of the RFFI (the project № 17-08-00849/17).

References

[1] Gavva L M and Endogur A I 2018 Statics and buckling problems of aircraft structurally-anisotropic composite panels with the influence of production technology *IOP Conference Series: Materials Science and Engineering* **312** 012009
[2] Setoodeh A R and Karami G 2003 *Compos. Struct.* **60** 245–53
[3] Gangadhara P B 2008 *Int. J. Mech. Sci.* **8** 1326–33
[4] Mittelstedt C and Schroder K U 2010 *Compos. Struct.* **92** 2830–44
[5] Yshii L N, Lucena Neto E, Monteiro F A C and Santana R C 2018 *Journal of Engineering Mechanics* **8** 04018061
[6] Ragb O and Matbuly M S 2017 *International Journal of Computational Methods in Engineering Science and Mechanics* **6** 292–301
[7] Castro S G P and Donadon M V 2017 *Composite Structures* **160** 232–47
[8] Shukla K K and Nath Y 2002 *Trans. ASME. J. Appl. Mech.* **5** 684–92
[9] Chen C S, Lin C Y and Chen R D 2011 *Int. J. Mech. Sci.* **53** 51–8
[10] Matsunaga H 2005 *Compos. Struct.* **68** 439–54
[11] Cetkovic M 2016 *Composite Structures* **142** 238–53
[12] Cetkovic M and Gyorgy L 2016 *Structural Integrity and Life* **1** 43–8
[13] Kettaf F Z, Benguediab M and Tounsi A 2015 *Periodica Polytechnica Mechanical Engineering* **4** 164–8
[14] Naik N S and Sayyad A S 2019 *Journal of Thermal Stresses* **5** 559–79
[15] Chen X, Dai S and Xu K 2001 *Ziran kexue ban* **41** 77–9 83
[16] Pandey R, Shukla K K and Jain A 2009 *Commun. Nonlinear Sci. and Numer. Simul.* **4** 1679–99
[17] Vescovini R and Dozio L 2015 *Compos. Struct.* **127** 35668
[18] Kazemi M 2015 *Arch. Appl. Mech.* **85** 1667–77
[19] Yeter E, Erklig A and Bulut M 2014 *Compos. Struct* **118** 19–27
[20] Abramovich H, Weller T and Bisagni C 2008 *J. of Aircraft* **45 (2)** 402–13
[21] Huang L, Sheikh A H, Ng C T and Griffith M C 2015 *Compos. Struct.* **122** 41–50
[22] Guo M W, Harik I E and Ren W X 2002 *Int. J. Solid and Struct.* **11** 3039–55
[23] Thankam V S, Singh G, Rao G V and Rath A K 2003 *Compos. Struct.* **59** 351–9
[24] Tenek L T 2001 *AIAA Journal* **39 (3)** 546–8
[25] Tran L V, Wahab M A and Kim S E 2017 *Composite Structures* **179** 35–9
[26] Kumar S, Kumar R, Mandal S and Ranjan A 2018 *International Journal of Civil Engineering and Technology* **6** 586–94
[27] Kumar S, Kumar R, Mandal S and Rahul A K 2018 *Open Civil Engineering Journal* **1** 468–80
[28] Zarei A and Khosravifard A 2019 *Composite Structures* **209** 206–18
[29] Castro S G P, Donadon M V and Guimaraes T A M 2019 *Composite Structures* **209** 67–78
[30] Falzon B G, Stevens K and Davies G O 2000 *Compos. A* **5** 459–68
[31] Park O, Haftka R T, Sakar B V, Starnes J H and Nagendra S 2001 *J. Aircraft* **2** 379–87
[32] Rouse M and Assadi M 2001 *J. Aircraft* **5** 950–5
[33] Ungbhakorn V and Singhatanagdid P 2003 *Compos. Struct.* **59** 455–65
[34] Baker D J 2000 *J. Aircraft* 1 **37** 138–43
[35] Zhao W, Xie Z, Wang X, Li X and Hao J 2019 *Mechanics of Advanced Materials and Structures* **3** 215–23
[36] Bai R, Bao S, Lei Z, Liu D and Yan C 2018 *Ocean Engineering* **160** 382–8
[37] Kumar S, Kumar R and Mandal S 2018 *International Journal of Civil Engineering and Technology* **6** 1324–32
[38] Sanchez M L, De Almeida S F M and Carrillo J 2017 *Revista Latinoamericana de Metalurgia y Materiales* **1** 45–9
[39] Firsanov V V and Gavva L M 2017 *Structural Mechanics of Engineering Constructions and Buildings* **4** 66–76

About the Steady Horizontal Circulation of Airship

V T Grumondz[1], N I Morozov[1] and Moung Htang Om[1]

[1] Moscow Aviation Institute, Moscow, Russia

E-mail: V.grumondz@gmail.com

Received xxxxxx
Accepted for publication xxxxxx
Published xxxxxx

Abstract

In this paper, the questions of the existence and calculation of the steady circulating motion of the airship in the horizontal plane are considered.

Keywords: airship, flight dynamics, steady motion, lateral motion, steady horizontal circulation

1. Introduction

In recent years, active work has been carried out all over the world on the design of airships and other aircraft, using, in addition to aerodynamic, the aerostatic principle [1–27]. In the design process, an important place belongs to the studies on the dynamics of the flight of an airship, which is very different from the dynamics of flights of aircraft heavier than air. However, until now, the problems of the dynamics of flight of an airship as a new type of aviation equipment remain completely unexplored. This article is devoted to one of the important problems of the airship's flight dynamics — the study of the existence and algorithms for calculating the parameters of the steady motions of the airship in the horizontal plane and is a natural continuation of the work [27] of one of the authors.

As noted in [27], the airship's flight dynamics has some similarities with the motion dynamics of an underwater vehicle of a similar shape, completely immersed in water during continuous flow. First of all, this is due to the great influence that the dynamics of the airship have on the value of the aerostatic force and its moments, as well as the magnitude of the attached masses, static moments and moments of inertia. In turn, this is due (as in the case of body motion in a fluid) the proximity of the average density of the airship to the density of the medium in which it moves. Based on these considerations, the differential equations of the spatial motion of the airship were obtained in [27], which we will use in this article.

2. Steady circulating airship motions in the horizontal plane

It's assumed that the airship is made according to the classical scheme, i.e. has a body in the form of an elongated body of revolution, a cross-shaped tail located in the rear, a nacelle with equipment or a crew, and two rotary screw propellers located symmetrically with respect to the longitudinal plane of symmetry. It will also be assumed that the airship body is an absolutely solid (non-deformable) body and changes in the mass, moments of inertia and attached masses of the airship during the considered period of flight time are negligible ($m = \text{const}$, $J = \text{const}$, $\lambda_{ik} = \text{const}$). As the beginning of the coordinate system associated with the airship $Oxyz$, we choose the center of volume of the airship body. This point will be called the pole of the airship. The coordinates of the center of mass of the airship in the system $Oxyz$ will be denoted as x_c, y_c, z_c, in the future it will be assumed that $z_c = 0$. It will also be assumed that the centrifugal moments of inertia J_{xz}, J_{yz} are negligible.

The general vector equations for the dynamics of the spatial motion of the airship, as in [27], can be described in the following form;

$$\frac{d\vec{Q}}{dt} + \vec{\omega} \times \vec{Q} = \vec{F};\ \frac{d\vec{L}}{dt} + \vec{\omega} \times \vec{L} + \vec{V} \times \vec{Q} = \vec{M}. \quad (1)$$

Where $\vec{Q}, \vec{L}$ — the main vector and the main moment of the total momentum of the airship and its ideal (not having viscosity) continuous medium; $\vec{F}, \vec{M}$ — the main vector and

IOP Conf. Series: Materials Science and Engineering **714** (2020) 012008 doi:10.1088/1757-899X/714/1/012008

the main moment of the external forces acting on the airship, with the exception of inertial forces, already taken into account in the left parts of these equations. The right-hand sides of this system are vectors

$$\bar{F} = \bar{G} + \bar{A} + \bar{P} + \bar{R}, \ \bar{M} = \bar{M}_G + \bar{M}_P + \bar{M}_R.$$

Where $\bar{G}$ — weight of the airship taking into account the carrier gas and air inside the hull; $\bar{A}$ — aerostatic force vector; $\bar{P}$ — vector of the resultant forces created by the propulsions; $\bar{R}$ — vector of resultant aerodynamic forces; $\bar{M}_G, \bar{M}_P, \bar{M}_R$ — the main moments of gravity, thrust propulsion and aerodynamic forces relative to the pole O. When recording the projections of the vectors $\bar{R}, \bar{M}$, the traditional dimensionless aerodynamic coefficients $c_x, c_y, c_z, m_x, m_y, m_z$ will be used. For the airships of the classical scheme, it is accepted to choose $l = U^{1/3}$ as a characteristic linear size, and as a characteristic area — value $S = U^{2/3}$.

It will be assumed that the volume of the instrument (cargo) nacelle and engines is small compared to the volume of the hull, so that the axis Ox is the axis of geometric symmetry of the entire airship. Then in the matrix of attached masses, only the following will be nonzero: $\lambda_{ii}, i = 1...6, \lambda_{26}, \lambda_{35}$.

System (1) must be supplemented with well-known equations of kinematic and geometric relationships, which are not given here.

Under the assumptions made, the system of equations for the steady horizontal circulation of an airship with a constant radius will have the form.

$$\begin{aligned}
&-mV_x\omega_y + (m+\lambda_{22})V_y\omega_x = (\rho g U - G)\cos\vartheta\sin\gamma + c_z qS;\\
&-my_c V_x\omega_y + my_c V_y\omega_x = Gy_c\cos\vartheta\sin\gamma + m_x qU;\\
&mx_c V_x\omega_y - (mx_c + \lambda_{26})V_y\omega_x = -Gx_c\cos\vartheta\sin\gamma + m_y qU; \qquad (2)\\
&V_x = V\cos\alpha\cos\beta;\ V_y = -V\sin\alpha\cos\beta;\ \omega_x = \omega_y\frac{tg\vartheta}{cod\gamma}.
\end{aligned}$$

For aerodynamic coefficients, the following structure will be taken:

$$\begin{aligned}
&c_z = c_z^{\beta}\beta + c_z^{\beta^2}|\beta|\beta + c_z^{\bar{\omega}_y}\bar{\omega}_y + c_z^{\beta\bar{\omega}_y}|\beta|\bar{\omega}_y + c_z^{\delta_H}\delta_H;\\
&m_x = m_x^{\beta}\beta + m_x^{\bar{\omega}_x}\bar{\omega}_x + m_x^{\delta_э}\delta_э; \qquad (3)\\
&m_y = m_y^{\beta}\beta + m_y^{\beta^2}|\beta|\beta + m_y^{\bar{\omega}_y}\bar{\omega}_y + m_y^{\beta\bar{\omega}_y}|\beta|\bar{\omega}_y + m_y^{\delta_H}\delta_H.
\end{aligned}$$

In system (2) $\beta, \bar{\omega}_y, \gamma, \delta_H, \delta_э$ are unknown; functions V, α, ϑ must be obtained in advance as a result of solving the equations of longitudinal steady motion of work [27]. Thus, two of the five unknowns must be specified. The sign β can be arbitrarily assigned at the beginning of the calculation, followed by verification of the assumption made. Note that the first and third equations (2) under assumption (3) can be considered independently of the second.

3. Algorithms for calculating horizontal circulation parameters

Let's consider some possible linear and nonlinear algorithms for calculating the parameters of the steady horizontal circulation of the airship and the corresponding balancing angles.

1. The angle β is small. Given ω_c, γ (or $\bar{\omega}_y, \gamma$). System (2) can be linearized. From the first and third equations, the balancing angles β, δ_H can be obtained, then from the second — the balancing angle $\delta_э$. These solutions are as follows:

$$\beta_* = \frac{1}{\kappa}\left[c_z^{\delta_H}q_3 - m_y^{\delta_H}q_1U^{\frac{1}{3}}\right];\ \delta_{H^*} = \frac{1}{\kappa}\left[m_y^{\beta}q_1U^{\frac{1}{3}} - c_z^{\beta}q_3\right];$$

$$\delta_{э^*} = \frac{1}{\kappa m_x^{\delta_э}}\left[c_z^{\delta_H}m_y^{\beta}q_2 + m_x^{\beta}m_y^{\delta_H}q_1U - c_z^{\delta_H}m_x^{\beta}q_3 - c_z^{\beta}m_y^{\delta_H}q_2\right],$$

where

$$\kappa = \frac{\rho V_*^2}{2}U\left[c_z^{\delta_H}m_y^{\beta} - c_z^{\beta}m_y^{\delta_H}\right]; \qquad (4)$$

$$\begin{aligned}
&q_1 = -V_*\omega_*\left[m + (m+\lambda_{22})\alpha_*^2\right] - (\rho g U - G)\gamma_* -\\
&-V_*\omega_*\frac{\rho U}{2}\left(c_z^{\bar{\omega}_y} + c_z^{\bar{\omega}_x}\alpha_*\right);\\
&q_2 = -V_*\omega_*\left[my_c\left(1+\alpha_*^2\right) + (\lambda_{26}+\lambda_{35})\alpha_*\right] -\\
&-Gy_c\gamma_* - V_*\omega_*\frac{\rho U^{\frac{4}{3}}}{2}\left(m_x^{\bar{\omega}_y} + m_x^{\bar{\omega}_x}\alpha_*\right);
\end{aligned}$$

In (4), it is possible to additionally take into account the smallness of the terms proportional α_*^2, in comparison with the terms containing α_* in the first and zero degrees, but for the future this is not significant. When $\gamma_* = 0$, balancing angles are independent of speed V_*.

2. The angle β is not small. System (2) is nonlinear. The following two possible algorithms for solving the problem of finding balancing values are possible:

a) Given ω_c, γ (or $\bar{\omega}_y, \gamma$). Imagine $\cos\beta \approx 1 - \frac{1}{2}\beta^2$ we obtain from (2) a system of the following structure:

$$\begin{aligned}
&g_{2i}\beta^2 + g_{1i}\beta + g_{0i} + g_i'\delta_H,\ i = 1,2,\\
&h_2\beta^2 + h_1\beta + h_0 + h'\delta_э = 0.
\end{aligned}$$

To obtain balancing values β, δ_H, the system of the first two equations $(i = 1,2)$ is solved. Then, substituting β those found in the third equation, find $\delta_э$. The expressions for the coefficients g_{ki}, h_k, g_i', h', as well as the final formulas for the solution, are given in view of the obviousness of the question;

b) Given γ, β. The system of the first and third equations (2) turns out to be linear in the variables $\bar{\omega}_y, \delta_H$ (or ω_c, δ_H),

IOP Conf. Series: Materials Science and Engineering **714** (2020) 012008 doi:10.1088/1757-899X/714/1/012008

so that it can be written their explicit expressions. It will done this for the case $\gamma = 0, |\alpha| \leq 3^{\circ}, |\vartheta| \leq 3^{\circ}, 0 < \beta < 30^{\circ}$, which is of greatest interest from the point of view of practical operation of the airship. It's obtained

$$\omega_y = \frac{1}{\Delta}(b_1 a_{22} - b_2 a_{12}),\ \bar{\omega}_y = \frac{1}{\Delta}(b_1 a_{22} - b_2 a_{12}), \qquad (5)$$

where

$$a_{11} = c_z^{\bar{\omega}_y} + c_z^{\beta\bar{\omega}_y}\beta + 2\mu\cos\beta;\ a_{12} = c_z^{\delta_H};\ \mu = \frac{m}{\rho U};$$

$$a_{21} = m_y^{\bar{\omega}_y} + m_y^{\beta\bar{\omega}_y}\beta - 2\mu\bar{x}_c\cos\beta;\ a_{22} = m_y^{\delta_H};$$

$$\bar{x}_c = x_c U^{-\frac{1}{3}};\ b_1 = -(c_z^{\beta}\beta + c_z^{\beta^2}\beta^2);$$

$$b_2 = -\left(m_y^{\beta}\beta + m_y^{\beta^2}\beta^2\right);\ \bar{y}_c = y_c U^{-\frac{1}{3}};$$

$$\Delta = (a_{11}a_{22} - a_{21}a_{12}).$$

Then, from the second equation (2) the required balancing angle of rudder deflection is found:

$$\delta_Э = -\frac{1}{m_x^{\delta_Э}}\left[\left(2\mu\bar{y}_c\cos\beta + m_x^{\bar{\omega}_x} tg\,\vartheta\right)\bar{\omega}_y + m_x^{\beta}\beta\right].$$

The radius of the steady circulation is

$$R_c = \bar{R}_c l = \frac{1}{\bar{\omega}_y}\cos\vartheta\cos\gamma U^{\frac{1}{3}}.$$

4. Conclusions

In the general case, an iterative algorithm for calculating the parameters of the steady circulation of an airship can be represented as follows:

1. The calculation of the parameters $\alpha_*^{(1)}, \delta_*^{(1)}, P_*^{(1)}$ of the longitudinal balancing in the first step for given V, φ, Θ and $c_x = c_{x0} = const$.

2. Calculation of balancing values $\beta^{(1)}, \delta_H^{(1)}, \delta_Э^{(1)}$ for any of the modifications of the above algorithms.

3. The calculation $c_x^{(2)} = c_{x0}^{(1)} + \Delta c_x(\alpha^{(1)}, \beta^{(1)}, \delta_H^{(1)}, \delta_Э^{(1)})$ in the second step.

4. Calculation $\alpha^{(2)}, \beta^{(2)}, P^{(2)}$ in the second step, etc.

With the help of (5), by varying β, it is possible to construct the so-called airship controllability diagram as the geometrical place of the points of the plane $(\bar{\omega}_y, \delta_H)$, to which steady circulating motions in the horizontal plane may correspond.

References

[1] Khoury G A and Gillett J D 1999 *Airship Technology* (Cambridge: Cambridge University Press)

[2] Khoury G A 1998 Review of unconventional design *24th Joint Annual Symposium of the Airship Association, June 26–28, Bedford* pp 21–42

[3] Nayler A 1998 Airship activity and development world-wide *24th Joint Annual Symposium of the Airship Association* pp 1–19

[4] De Laurier J D 1972 *J. Aircraft* **9** 646–51

[5] De Laurier J D, Lowe J D, Uffen D R and Amann J H 1986 Research on airship flight dynamics at the university of Toronto *Conf. on "Airship design and operation – present and future"* (Royal Aeronautical Society)

[6] Fichtl G H Problems in simulation of atmospheric boundari layer flows *AGARD CP-140*

[7] Houbolt J C 1973 Survey on effect of surface wind on aircraft design and operation and recomendations for needed wind research *NASA CR-2360*

[8] Ramos J, Bueno S, Maeta S, de Paiva E, Asamuma K, Nascimiento L, Bergerman M, Elfes A, Antonio J and Beiral R 1998 Project AURORA. *The 24th Joint Annual Symposium organized by the Airship Association, 26–28 June 1998, Bedford, England* pp 91–104

[9] Oswald W B 1932 *The Transverse Force Distribution on Ellipsoidal and Nearly Ellipsoidal Bodies Moving in an Arbitrary Potential Flow. Thesis* (Calif. Institute of Technology)

[10] Bicho E, Moreira A, Carvalheira M and Erlhagen W *2005 Autonomus Flight Trajectory Generation via Attractor Dynamics International Conference on Intelligent Robots and Systems IEEE/RSJ* pp 615–21

[11] Bestaoui S Y 2012 Lighter than Air Robots *Intelligent Systems, Control and Automation* (Springer Science+Business Media BV)

[12] Knoblock F D and Troller Th 1935 *Tests on Effect of Sidewind on the Ground-Hanling of Airships* (Daniel Guggenheim Airship Institute)

[13] Ashford R L 1986 Military & civilian applications for tethered aerostats *Conference on Sounding Rockets, Ballons & related Space Systems, AIAA 7th, Oct, Ocean City Md, Collection of Technical Papers, NewYork NY* pp 28–34

[14] Thompson N 1972 *Quart. J. R. Met. Soc.* **98** 745–62

[15] Rice C L, Korn A O, McPhetres G H and Hunsinger F V Some unique applications for tethered balloon systems *AIAA* **81**

[16] De Laurier J D 1977 *Prediction of Tethered-Aerostat Response to Atmospheric Turbulence. Engineering Notes* (Amsterdam, Netherlands)

[17] Jones S P and Krausman J A 1982 *Journal of Aircraft* **19** 679–86

[18] Jones S P 1989 Prediction of aerostat & airship mooring mast loads by nonlinear dynamic simulation *AIAA Lighter-Than-Air Systems Technology Conference, 8th, Jacksonville, Fl, Oct. 5–7* Technical Papers, Washington, DC, AIAA

[19] Grumondz V and Matus V 1998 Dynamics of an airship connected to earth with a rope *The 24th Joint Annual Symposium organized by the Airship Association, 26-28 June, Bedford, England* pp 73–80

[20] Zasolov R 1985 *TsAGI* **2268** 31–46

[21] Arie M 1986 *Airships* (Kiev: Naukova dumka)

[22] Boiko Y 2001 *Aeronautics* (Moscow: MGUP)

[23] Kirilin A 2013 *Airsips* (Moscow: MAI-Print)

[24] Grumondz V ans Rudnev V 1989 Airship established horizontal circulation *22-nd Tsiolkovsky readings* (Moscow, IIET of the the USSR Academy of science) pp 103–7

[25] Grumondz V and Matus V 1999 *MAI Vestnik* **2** 60–6
[26] Loytsansky L 1970 *Liquid and gas mechanics* (Moscow: Nauka)
[27] Grumondz V 2019 Airship Balancing and Stability at Longitudinal Established Motion. *Workshop on Materials and Engineering in Aeronautics. IOP Publishing. IOP Conf.Series: Materials Science and Engineering* **476** 012013

Design of an Experimental Test Bench for a UAV Type Convertiplane

S Jatsun[1], O Emelyanova[1] and A S Martinez Leon[1]

[1] Southwest State University, Kursk, Russia

E-mail: teormeh@inbox.ru

Received xxxxxx
Accepted for publication xxxxxx
Published xxxxxx

Abstract

This paper deals with the design of a test bench for an unnamed aerial vehicle UAV type convertiplane with controlled electric drives and tilt rotors (propellers) executed on a quadcopter pattern. This experimental test bench provides an intermediate step between simulation process and real time flight of the UAV, it allows to test its and evaluate solutions in safe and controllable environment conditions. Thus, the development of methods of state estimation and control UAV becomes possible, in addition the developed system can be used for teaching about the basis of automatic control in multirotor aerial vehicles. The test bench presented in this paper contains a convertiplane type quadcopter attached on the base through a universal joint which let the UAV be able to rotate around plane axes x and y, while the vertical axis z is fixed, in turn this part of the system is fixed through a bar connected with a linear guideway block that moves over a linear guide which depends of the generated lifting force by the UAV. To determinate the travelled road of the system and the different parameters of the behavior of the convertiplane it has been also implemented a measuring system.

Keywords: test bench platform, convertiplane, mathematical modelling of UAVs, UAV control systems, UAV trajectory planning

1. Introduction

Nowadays small-sized unmanned aerial vehicles (UAVs) represent a suitable alternative for monitoring, ground mapping, agricultural and environmental preservation practices, fire detection, transmission power line infra-red supervision, etc. in large, dangerous or remote areas. UAV systems can be equipped with a wide variety of sensors due to the given task [1–3, 5, 6, 9]. In addition, the use of modern manufacturing capabilities and element base, the availability of new composite materials have led to production of aerial vehicles with lightweight and strong hulls; the development of micro-electronic components, such as microcontrollers, sensors, receivers, drivers, miniature-type cameras, etc. helped to create high-precision navigating and flight stabilization systems for aerial vehicles, as well as digital data transmission over long distances [2, 4, 7, 8, 10]. Control techniques of multirotor aerial vehicles is a particularly discussed topic in several works of Russian and foreign researchers.

Currently the analysis of existing models of autonomous multi-rotor highly manoeuvrable small unmanned aerial vehicles (UAV) weighing up to 5 kg, that can carry payload, is the most promising and needed field of research for both the military and civil spheres. Modern UAV is a flying laboratory, which includes an aerial vehicle and a hardware-software complex that allows operating the device in manual, semiautomatic or automatic mode, as well as receiving, storing and displaying the telemetry data received from the UAV.

After concluding a UAV project, it is indispensable to perform tests and evaluate its controlled response, such as check up its control loops to check if it reached the project requirements.

There are two options for checking the proper functioning of a drone: real-time outdoor test and test using a platform. Especially when we talk about the development of completely autonomous aerial systems, as well as non-traditional configuration of drones, the carrying out of outdoor tests generates a high risk on the UAV, which can lose control or bump into an obstacle at any time, etc. Thus, it is advisable to consider the use of safe and reliable experimental platforms, it allows to test it and evaluate solutions in safe and controllable environment conditions.

There are different concepts of test benches which vary from the simplest approaches, where the UAV is fixed to a joint with wires [4] or rotates around single fixed axis [5]. The most sophisticated devices deliver more than 3 DOF [2]. The aim of this paper is to present a useful, versatile and low-cost solution for testing and tuning the flight controller before the phase of flight tests.

This paper is organized as follows: Section 2 presents a full description of the kinematics and dynamics of the developed UAV type convertiplane executed on a quadcopter pattern used in this work with; Section 3 shows the designed test bench platform in this work, with all its constructive characteristics; Section 4 presents the simulation results obtained in this work; Section 5 concludes this purpose, such as some notes about future work.

2. Mathematical modelling of a UAV type convertiplane

The scheme of the quadcopter convertiplane is provided in figure 1. Consider the motion of the convertiplane in a fixed cartesian coordinate system *OXY*. Let us assume that CX_1Y_1, $A_ix_iy_i$ $(i=1–4)$ are local coordinate systems, passing through the center of mass of the body of the apparatus C and the centers of mass A_i of *i-th* electric drive with propellers, hereinafter referred to as the rotor. The orientation of the body in the air is set by the aircraft yaw angles ψ, pitch θ and tilt φ angles. However, from the stability point of view, it is of interest to study the pitch of the device relative to its transverse axis CZ_1 at an angle θ when the thrust vectors of rotary drives 1,2 at an angle β, which is carried out in a plane parallel to the plane CX_1Y_1. Angle β can vary from 0 to 90°.

Now we will denote by the symbols (0) and (1) the vectors defined in the coordinate systems *OXYZ* and *CX1Y1Z1*, respectively. The position of the center of mass of the convertiplane is given by the vector $\bar{r}_{OC}^{(0)} = [X, Y, Z]^T$, and its orientation in space by the vector $\bar{\alpha} = [\phi, \psi, \theta]^T$ [4]. Thus, the vector $\bar{r}^{(0)} = [\bar{r}_{OC}^{(0)}, \bar{\alpha}]^T$ fully defines the spatial location of the convertiplane.

The forces applied to the convertiplane, which can be divided into forces that are set in absolute system of coordinates *OXY*: gravity force $\bar{G}$, applied to the center of mass *C* and forces, defined in the movable coordinate system CX_1Y_1: the forces generated by the propellers $\bar{F}_i$, applied at the center of mass of the rotors $\bar{A}_i$; aerodynamic forces acting on the wing surface $Q_{1,2}$, called the force of drag, directed to the opposite side of the axis CX_1 and Q_ϕ force acting on the fuselage and lifting forces of the wing — $\bar{P}_{1,2}$ and fuselage $\bar{P}_\phi$ [6, 8].

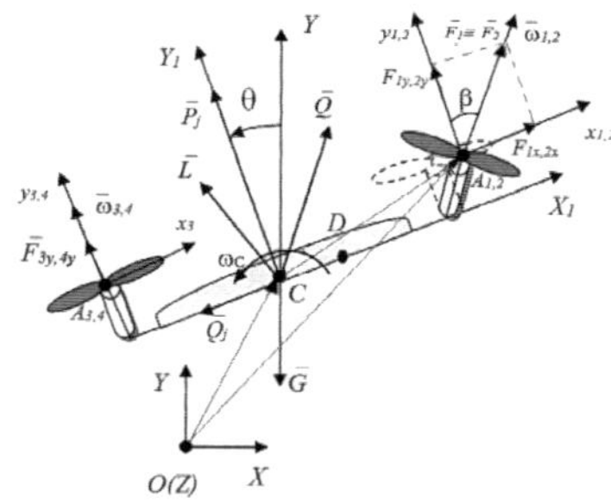

Figure 1. Calculation scheme for the study of movements relative to the transverse axis CZ_1): $\bar{Q}, \bar{L}$ — vector of the quantity of motion and the kinetic moment; $\boldsymbol{\omega_i}$— vector of the angular velocity of rotation of the *i*-th rotor; ω_C — angular velocity of rotation of the convertiplane body.

The moments of traction forces can be presented in the following way:

$$\bar{M}_i = \bar{r}_{CA_i} \cdot \bar{F}_i^{(1)} \quad (1)$$

and the thrust vector:

$$\bar{F}_i^{(1)} = \bar{b}_i^T \bar{\omega}_i \, |\bar{\omega}_i| \quad (2)$$

where $\bar{b}_i = (b_x \quad b_y \quad b_z)^T$ — aerodynamic coefficient, the module of which is $b_i = \frac{c_T \rho S_H R^2}{2}$ [6, 8, 12].

The thrust vectors of the propellers, wing lift force and drag force in the zero-coordinate system can be determined in accordance with the following formulas [7, 8, 10]:

$$\bar{F}_i^{(0)} = T_{10} \cdot \bar{F}_i^{(1)}, \bar{P}_\phi^{(0)} = T_{10} \cdot \bar{P}_\phi^{(1)}, \quad \bar{Q}_\phi^{(0)} = T_{10} \cdot \bar{Q}_\phi^{(1)}$$
$$\bar{P}_{1,2}^{(0)} = T_{10} \cdot \bar{P}_{1,2}^{(1)} \quad , \quad \bar{Q}_{1,2}^{(0)} = T_{10} \cdot \bar{Q}_{1,2}^{(1)} \quad (3)$$

where T_{10} is a transition matrix from $^{(1)}$ to $^{(0)}$ coordinate systems has the following form:

$$T_{10} = \begin{bmatrix} \cos\theta & -\sin\theta & 0 \\ \sin\theta & \cos\theta & 0 \\ 0 & 0 & 1 \end{bmatrix} \quad (4)$$

The forces acting on the wing surface depend on the design and nature of the load. Thus, the position of the center of the aerodynamic load pressure depends on the flight mode. In accordance with the theorems of mechanics, it is possible to bring an arbitrary system of forces to the center of mass, i.e. $\{\bar{P}_{1,2,\phi}, \ \bar{Q}_{1,2,\phi}\} \sim \{\bar{P}_j, \ \bar{Q}_j, \ \bar{M}_{jC}\}$, where j =1–4.

The modules of the wing lift force $P_j^{(1)}$ and drag force $Q_j^{(1)}$ will determine by the following formulas [7, 8]:

$$P_j^{(1)} = \frac{c_L \rho S V^2}{2} = \mu \bar{V} |\bar{V}|,\ Q_j^{(1)} = \frac{c_D \rho S V^2}{2} = k \bar{V} |\bar{V}|, \quad (5)$$

where ρ is the air density, c_L is the lift coefficient, c_D is the coefficient of drag, which depends on the geometry of the wing, S — wing surface, V — speed of the incoming airflow.

The system of differential equations describing the change in the generalized coordinates, determining the position of the convertiplane center of mass in space under the action of the propeller thrust forces, lifting force, drag forces and weight forces can be obtained from the general theorems of dynamics [4, 11, 12]:

$$\frac{d\bar{Q}}{dt} = \sum \bar{F}_i^e, \quad (6)$$

$$\frac{d\bar{L}}{dt} = \frac{\tilde{d}\bar{L}}{dt} + (\bar{\omega}_C \times \bar{L}) = \sum \bar{M}_C^e. \quad (7)$$

Vectors $\bar{Q}$ and $\bar{L}$ for a mechanical system consisting of a body and 4 propellers will have the form:

$$\bar{Q} = \sum_{i=1}^{5} m_i \bar{v}_i;\ \bar{L} = \bar{L}_C + \sum_{i=1}^{4} I_i \bar{\omega}_i \quad (8)$$

where I_i — inertia tensor.

Then, we present the vector product of equation (7) in matrix form [12]:

$$(\bar{\omega}_C \times \bar{L}) = \lambda(\bar{\omega}_C) \cdot \bar{L} =$$

$$\begin{vmatrix} 0 & -\omega_{CZ_1} & \omega_{CY_1} \\ \omega_{CZ_1} & 0 & -\omega_{CX_1} \\ -\omega_{CY_1} & \omega_{CX_1} & 0 \end{vmatrix} \cdot \begin{vmatrix} L_{X_1} \\ L_{Y_1} \\ L_{Z_1} \end{vmatrix} = \begin{vmatrix} L_{Z_1}\omega_{CY_1} - L_{Y_1}\omega_{CZ_1} \\ L_{X_1}\omega_{CZ_1} - L_{Z_1}\omega_{CX_1} \\ L_{Y_1}\omega_{CX_1} - L_{X_1}\omega_{CY_1} \end{vmatrix} \quad (9)$$

In our case $\omega_{CZ_1} = \dot{\theta}$, $\omega_{CX_1} = 0$, $\omega_{CY_1} = 0$.

Then, taking into account (9), equations (6) and (7) will take the form:

$$\dot{\bar{Q}} = \bar{F}_i^e$$

$$\dot{\bar{L}} = \tilde{\dot{\bar{L}}} + \lambda(\bar{\omega}_C)\bar{L} = \sum \bar{M}_C^e \quad (10)$$

Next, we consider the linear motion of the convertiplane, so the moment of traction forces of the propellers relative to the axis CZ_1 has the form:

$$\sum_{j=1}^{4} M_{jZ_1}^e = (F_2 \cos\beta - F_4) b_z + (F_1 \cos\beta - F_3) b_z =$$

$$= b_{z1}\left((\omega_2^2 + \omega_1^2)\cos\beta + b_{z2}(-\omega_3^2 - \omega_4^2)\right) + \alpha P_j^{(0)} (\dot{X})^2 \quad (11)$$

or, taking into account the propeller drives [11, 12, 16–18], we obtain a system of nonlinear differential equations describing the plane motion of the convertiplane in the longitudinal plane:

$$\begin{cases} \ddot{X}(t) = \frac{1}{M}\left(\sum_1^4 F_{ix}^{(0)} + \sum_1^3 Q_{jx}^{(0)}\right) \\ \ddot{Y}(t) = \frac{1}{M}\left(\sum_1^4 F_{iy}^{(0)} - Mg + \sum_1^3 P_{jy}^{(0)}\right) \\ \ddot{\theta} = \sum_{j=1}^{4} \frac{M_{jZ_1}^e}{J_{Z_1}} \\ \dot{\omega}_i = f(U_i(\Delta)), \quad i = 1-4 \end{cases} \quad (12)$$

where $M = \sum M_k$ — mass of the convertiplane, $U_i = U_i(\Delta)$ — control voltages applied to electric motors, depending on the magnitude of the error $\Delta = (\Delta X, \Delta Y, \Delta Z)^T$, α — the arm of the wing lifting force.

$$U_i(\Delta) = k_p(\Delta) + k_d \frac{d\Delta}{dt} \quad (13)$$

where k_p, k_d — coefficients respectively proportional and differential controller.

Deviations of the real positions of the center of mass and the pitch angle θ of the convertiplane lead to errors and require corrective action:

$$\begin{aligned} X(t) - X^*(t) &= \Delta X \\ Y(t) - Y^*(t) &= \Delta Y \\ \theta - \theta^* &= \Delta\theta \end{aligned} \quad (14)$$

In general conditions the system (12) is nonlinear and its solution is convenient to perform it using numerical methods.

3. Development of a test bench platform for a UAV type convertiplane

In this study, a test platform for rotary wing UAVs with multi-rotors is presented. The proposed test platform draft done by SolidWorks is presented in figures 2a and 2b. As can be seen from this figure, the UAV that is hanged on to the test platform and can move easily at all axes x and y, also the convertiplane is able to realize a circular motion trajectory. By using this platform, the pre-flight tests, such as pitch, roll, yaw and elevation will be realized.

In order to present the results obtained from the test bench platform is used a Bluetooth wireless connection, the operator receives the information to the station using a computer where the information is plotted.

The developed test bench platform is presented in figures 2a and 2b and contains a support base 1, on which a linear guide 2 is fixed, in turn on the linear guide 2 a platform 3 is movably mounted. On the platform 3 is fixed a universal joint 4 (2DOF), it allows the rotation of the convertiplane 5 in the plane X,Y. A cable 6 is attached to the platform 3, which is connected to the load 9 through a system of blocks 7 and 8 mounted on the vertical linear guide. At the ends of the linear guide 2 a pair of bearings are installed. A measuring system is installed to determine the kinematic characteristics (not shown).

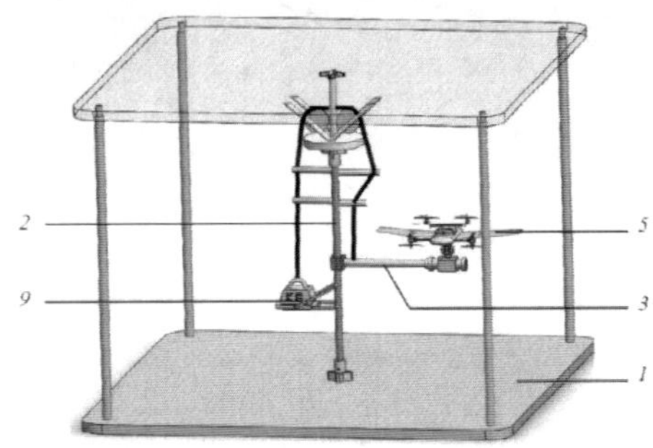

Figure 2a. Test bench platform for a UAV type convertiplane isometric view on software Solidworks: 1 — support base, 2 — linear guide, 3 — platform, 5 — convertiplane, 9 — load.

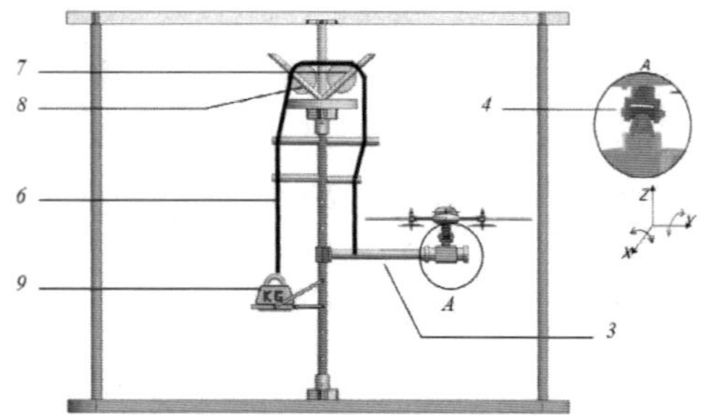

Figure 2b. Test bench platform for a UAV type convertiplane frontal view on software Solidworks: 4 — universal joint (2 DOF), 6 — cable, 7, 8 — system of blocks, 9 — load.

The test bench platform works as follows. The convertiplane 5 is fixed motionless on a two-axis universal joint 4, which is mounted on a platform 3. Through a cable 6 the load 9 is suspended such as the convertiplane 5 is in a certain equilibrium position. The system is turned on, the measuring system installed on-board the convertiplane 5 as in the bench platform are tested. The two-axis universal joint 4 allows the convertiplane to rotate in the horizontal plane X,Y. The communication and data transmission is carried out via Bluetooth. According to the given algorithm, the convertiplane 5 performs various manoeuvres in order to determine the created lift force, the behaviour during a circular trajectory checking the stability of the system when the UAV turns in the horizontal plane relative to the axes x and y. Thus, the test bench platform allows determine the kinematic characteristics of the UAV system.

4. Evaluation

In order to prove the concept of the presented test bench platform, few experiments were conducted. The objective of the tests was to confirm the proper operation of the presented system in development of flight controller.

It has been established that the process of modeling is divided in 3 stages: I — take off, II — trajectory, III — landing.

The first step for system modeling is the definition of the initial conditions to be used. For vertical flight we considered the initial coordinates of the UAV, which are given by the form:

$$X = X_0, Y = Y_0, Z = Z(t) \tag{15}$$

where X,Y,Z — given coordinates, X_0, Y_0 — initial coordinates for X and Y, $Z(t) = at^3 + bt^2 + ct + d$ — function used to describe the lift of the convertiplane.

To determine the value of the constants used in formula 15 is necessary to establish the initial conditions of the system as shown in table 1.

Table 1. Parameters for the lift and circular trajectory of the convertiplane

	Time [s]	Position [m]	Velocity [m/s]
Initial conditions	$t_0 = 0$	$Z_0 = 0$	$\dot{Z}_0 = 0$
Final conditions	$t_1 = t_1$	$Z_0 = H$	$\dot{Z}_1 = 0$

Then, the next step is about the modeling of the circular trajectory of the convertiplane. For circular trajectory we considered the following coordinates, which are given by the form:

$$X = a\sin(\omega t), Y = b\sin(\omega t), Z = const \tag{16}$$

where X,Y,Z — given coordinates; a, b — coefficients; ω — frequency; t — time

To determine the value of the constants used in formula 16 is necessary to establish the initial conditions of the system as shown in table 2.

Table 2. Parameters for the circular trajectory of the convertiplane

	Time [s]	Position [m]	Velocity [m/s]
Initial conditions	$t_1 = t_1$	$Z_1 = H$	$\dot{Z}_1 = 0$
Final conditions	$t_2 = t_2$	$Z_2 = H$	$\dot{Z}_2 = 0$

According to the geometrical characteristics of the developed test bench platform for a convertiplane, the maximum possible height value is 1 meter.

The results obtained from the modeling of the system are shown in figures 3–6.

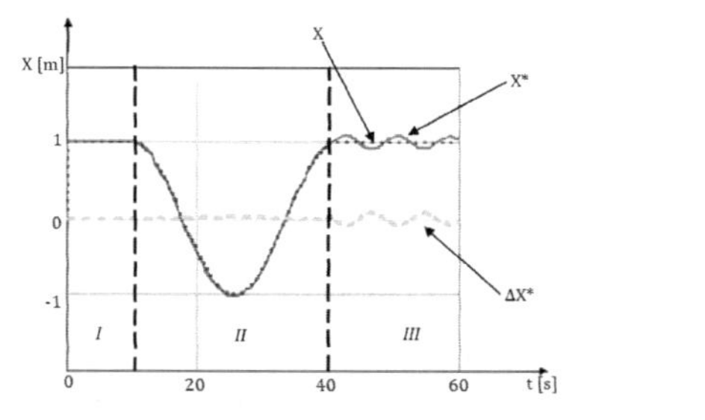

Figure 3. Movement of the convertiplane along x-axis: X — given trajectory, X^* — obtained trajectory, ΔX — error between X and X^*.

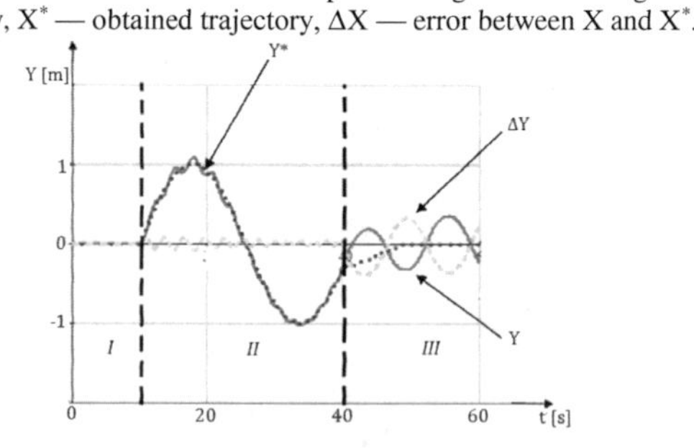

Figure 4. Movement of the convertiplane along y-axis: Y — given trajectory, Y^* — obtained trajectory, ΔY — error between Y and Y^*.

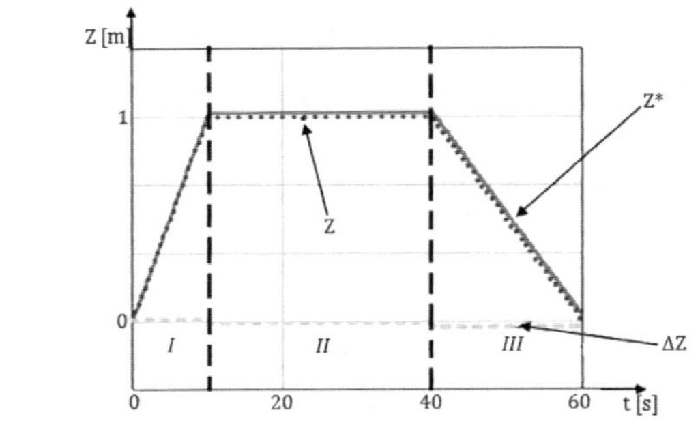

Figure 5. Movement of the convertiplane along z-axis: Z — given trajectory, Z* — obtained trajectory, ΔZ — error between Z and Z*.

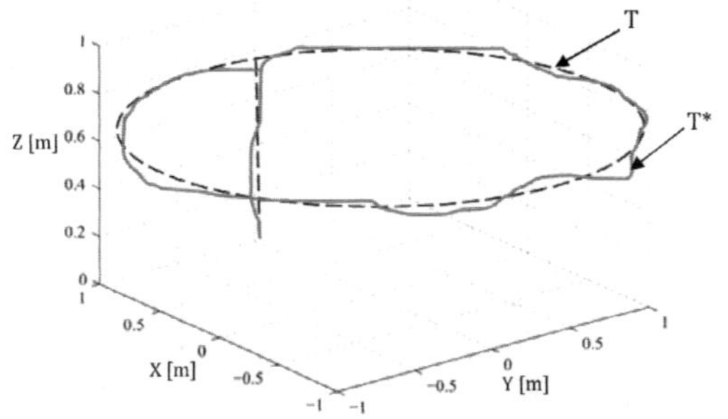

Figure 6. Three-dimensional modeling of the trajectory of the convertiplane: T — given trajectory, T* — obtained trajectory.

5. Conclusions

In this paper a test bench platform for an unnamed aerial vehicle UAV type convertiplane with controlled electric drives and tilt rotors (propellers) executed on a quadcopter pattern has been designed. Also a mathematical model of a UAV type convertiplane has been developed. The obtained results in real time flight of the UAV correspond to the results obtained through simulation process. A control method for the movement of the convertiplane, including the assignment of an arbitrary trajectory in form of functions has been proposed. Thus, the use of the developed test bench platform has made possible to test the designed UAV in a safe and controllable environment conditions. Further, there will be improved the physical model the test bench platform with the aim to positioning it as a research tool, as well as a teaching tool for the development of control systems and design in general of UAVs.

References

[1] Osedo A et al. 2015 Flight control systems of a quad tilt rotor UAV for a large attitude change *IEEE Robotics and Automation (ICRA) Conference* pp 2326–31

[2] Salazar-Cruz S et al. 2008 *IEEE Transactions on Aerospace and Electronic Systems Conference* **2** 783–94

[3] Emelyanova O V et al. 2017 The synthesis of electric drives characteristics of the UAV of convertiplane-tricopter type. *2016 Workshop on Contemporary materials and technologies in the aviation industry (CMTAI2016) MATEC Web of Conferences* **99** pp 249–63

[4] Jatsun S F et al. 2014 Mathematical model of the quadrotor type unmanned aerial vehicle with neurocontroller *Proceedings of the 18th International Conference on Circuits (CSCC '14) & Proceedings of the 2014 International Conference on Mechatronics and Robotics, Structural Analysis (MEROSTA 2014) Advances in Robotics, Mechatronics and Circuits* pp 46–50

[5] Jatsun S et al. 2017 Control flight of a UAV type tricopter with fuzzy logic controller *IEEE XIII International scientific and technical conference "Dynamics of Systems, Mechanisms and Machines"* pp 1–5

[6] Jatsun S et al. 2019 Issledovanie kolebanij konvertoplana v vertikal'noj prodol'noj ploskosti *Proceedings of Southwest State University* **23 4** pp 42–56

[7] Emelyanova O V et al. 2017 Sintez parametrov ehlektroprivodov BPLA tipa konvertoplan-trikopter *IV International School-Conference for Young Scientists "Nonlinear Dynamics of Machines" Mechanical Engineering Research Institute of the Russian Academy of Sciences* pp 239–49

[8] Efremov A V et al. 2011 *Dinamika poleta: uchebnik dlya studentov vysshih uchebnyh zavedenij* (Moscow: EDP Mashinostroenie)

[9] Martinez Leon A S 2015 Razrabotka konstrukcii konvertoplana *2nd International Scientific and Practical Conference 2, Southwest State University* pp 265–8

[10] Jatsun S et al. 2020 Investigation of oscillations of a quadcopter convertiplane in transient mode in the vertical longitudinal plane *Proceedings of 14th International Conference on Electromechanics and Robotics "Zavalishin's Readings"* (Singapore: Springer) pp 345–58

[11] Pavlovsky V E et al. 2015 Matematicheskoye modelirovaniye robota s peremennym vektorom tyagi *Bespilotnyye transportnyye sredstva s elementami iskusstvennogo intellekta (BTS-II-2015), Russian Association for Artificial Intelligence* pp 99–106

[12] Pavlovsky V E et al 2014 *Robototekhnika i tekhnicheskaya kibernetika* **4** 49–57

Modern Diagnostics of Aircraft Gas Turbine Engines

A S Komshin[1] and V I Pronyakin[1]

[1] Bauman Moscow State Technical University, Moscow, Russia

E-mail: VIP-U@yandex.ru

Received xxxxxx
Accepted for publication xxxxxx
Published xxxxxx

Abstract

The article deals with the assessment of the technical condition of gas turbine engines. The analysis of the application of traditional approaches to the assessment of the technical condition of gas turbine engines. The technique of obtaining parameters of low-frequency torsional vibrations of the shaft on the basis of precision chronometric measurements is presented.

Keywords: control, phase-chronometric method, gas turbine engine, measuring technology, construction material, diagnostics, predictive analytics

1. Introduction

Currently, the problem of predictive monitoring of the technical state of aircraft gas turbine engines (GTE) in order to improve reliability and emergency protection is becoming increasingly important [1].

There are different approaches to diagnosing the functioning of gas turbine engines. The laser-based experimental methods has created an unprecedented shift in turbulent combustion research [2–6].

The development and application of predictive maintenance concepts for several types of assets, following two approaches: (1) detection and prediction of failures based on (real-time) monitoring the health or condition of the systems, and (2) prediction of failures (prognostics) using physical failure models and monitoring of loads or usage [7].

The practical implementation of this approaches is discussed by showing a number of case studies in different sectors of industry [8–11].

An alternative methods of predictive Analytics and diagnostics of gas turbine engines can be the introduction of a state space model, embedded with particle filtration in a set of data on the degradation of a gas turbine [12–17].

Operation of the gas turbine engine is associated with a technical risk of malfunctions, the main ones are

- wear parts of the flowing part,
- mechanical damage of parts of the flowing part,
- deformation and destruction of parts and assemblies,
- violation of the settings in the GTD systems,
- violation of operating modes of units, systems and the engine as a whole.

Experience in the operation of GTE shows that (65÷70)% of all flight accidents associated with the accumulation of damage during operation to a critical level, after which there is a failure of a single unit or the entire engine. However, as a rule, this gradual accumulation of damage is not accompanied by any distinct features recorded by traditional approaches.

Another important task is the emergency protection and monitoring of the technical condition of the GTE in real time. It is possible to provide reliable emergency protection only by carrying out systematic measurement and computational monitoring of its technical condition.

Among the primary converters of physical quantities that characterize the processes in the GTE, are currently used

- induction and piezo metric sensors for measuring vibration parameters;

- microphones based on electrical and piezoelectric effects for measuring acoustic vibrations;
- strain gauges for measurement of constant and variable strain;
- resistance thermometers — for high temperature measurement;
- speed measuring instruments;
- indicators of the presence of chips in oil and metal particles in gases.

All the measuring instruments listed here have a level of relative errors not lower than (0.1=0.01)%.

However, such traditional approaches cannot effectively solve the problems of diagnostics and emergency protection of gas turbine engines. Experience has shown that they do not fully meet the diagnostic requirements of the engines currently in operation. Obviously, due to not enough resolution and the sensitivity, they cannot provide a diagnosis for promising gas turbine engines, namely:

- newly developed engines, where the necessary detailed information retrieval with the aim of testing new nodes;
- engines with higher speed and counter-rotation of the rotors;
- micro-turbojet, which have a speed of up to 2000 rpm.

Therefore, it seems necessary:

- to build an updated multi-factor mathematical models of the functioning of the GTE;
- to consider the possibility of improving the metrological characteristics of the GTE control and diagnostic system sensors, based on the achievements of Metrology and measuring equipment;
- to introduce compact built-in phase-chronometric systems [18, 19] into the onboard control and diagnostic equipment to ensure prompt receipt of measurement information on the parameters of rotating running parts of the GTE with relative error no worse than (10-3-10-4)%;
- identification of mathematical models of functioning and diagnostics GTE.

One of the main diagnostic factors in the operation of the engine as a rotary machine, containing information about its technical condition, are vibration and small fluctuations in the period of rotation of the rotors within the turnover. Spectra of vibrations and variations of the rotation period of the rotors of a serviceable engine are repeated in a certain tolerance field with multiple registrations during the long-term operation of the engine and change with damage to its elements of the structure.

The vibration spectra of the running engine are determined by the speed of the rotor; the frequency of the blades; the speed of the structural elements of the bearing and their condition; the speed of the gears, gears, elements of the units installed on the GTE (units, pumps, etc.) and their condition; the frequency of the natural vibrations of the elements of the engine design and their technical condition; the frequency of acoustic excitations; turbulence of the gas-air flow.

In flight, increased levels of vibration and fluctuations in the period of rotation of the engine rotors can be a consequence of:

- increases of unbalance of rotors;
- destruction of parts of the rotor;
- changes in the stiffness of the supporting part of the rotor-stator system;
- touches of rotor parts and stator parts;
- resonant vibrations of the elements of the aircraft structure;
- violations of concentricity of the shaft and bearing housing;
- oscillation of the rotor in the axial direction;
- self-oscillation of structural elements of engine systems;
- acoustic vibrations;
- withdrawal of the aircraft on the flight conditions, adverse vibration condition GTE.

The increased vibrations are the consequence of such damage, as

- cracks in the diaphragms, power racks and power elements of the housings;
- one-way radial wear of the treadmill of the inner ring of the rolling bearing and its destruction due to large radial loads;
- the destruction of the separators of rolling bearings from large relative displacements of the separator in the bearing caused by fluctuations in the rotor;
- teasers at the ends of the working blades and at the tops of labyrinth seals;
- residual deformation of the shaft during bending;
- destruction of the rotor pin connections.

Increased vibration levels and variations in the rotation period of the motor rotors are an important diagnostic feature for the following defects:

- breakage of compressor and turbine blades;
- damage of rotor blades of the foreign object;
- thermal imbalance;
- changing the rigidity of the rotor due to violations of the attaching discs of the compressor rotor;
- ingress of liquid (oil, water) into the rotor cavity, etc.

When the structural elements of the rotor-housing system are damaged, unbalance of the rotors occurs. The consequence is the appearance of harmonics with multiple and fractional frequencies f = ifp, where i=1, 2, 3... m are the numbers of multiple harmonics. The appearance of fractional harmonics (subharmonic oscillations) i=1.2/1.3, 1.4/possibly at/speeds ... of rotation equal to if CR where CR= TCR, 60 where= TCR/critical, the rotor speed. Since this speed is not included in the operating frequency range of rotation of the rotor, the presence of such components in the spectrum of the vibration will indicate a breach of the normal tion of the functioning of the CCD. The high-speed rotor may have components with harmonics i=3/2, 5/2 ... due to the peculiarities of the rotor axis movement in the bearing gaps.

Changes in the stiffness of the supporting part of the rotor-stator system can occur due to violations of the technical requirements of the engine installation on the summer-telny device or due to deformation or corrosion of the supports.

Vibrations of the rotor in the axial direction occur due to vibration combustion in the main and afterburner combustion chambers. Such oscillations are dangerous because they create conditions for excitation of compressor and turbine disks and acceleration of damages in them to a critical level.

Acoustic vibrations are caused by noise generated by the input device, compressor, turbine, combustion chamber and jet nozzle.

Violations of the concentricity of the shaft and the bearing housing during operation is possible as a result of deformation of the engine housing in extreme flight conditions with overloads exceeding the permissible ones.

Currently in operation is applied a method of estimation of technical condition vibration level and the rotation speed of the rotors. The method is based on the comparison of the current value of the vibration parameter with the maximum permissible (according to the scheme "good — not good"). It should be able to detect cases of exceeding the current value of the horn level of vibration. The disadvantage of the method is that the amplitude of the vibration signal may not reach the set limits. In addition, some damages may reduce the vibration level.

For early detection of emerging and gradually accumulating defects require a high sensitivity of measuring instruments and the selection of weak signals, carrying information on the background of intense noise. Traditional vibroacoustic method does not provide an effective solution to this problem. Therefore, it is important to use the possibilities of modern means and methods of precision chronometry [19, 20]. They are able to detect small variations in the period of rotation of the rotor within one revolution. For example, to obtain the parameters of torsional vibrations of shafts. It can be expected that phase-chronometric diagnostic systems with a higher metrological level will provide new information compared to traditional vibrodiagnostics.

2. Method of determination of low-frequency torsional vibrations of shafts.

A rotating shaft functioning GTE can perform torsional vibrations. Determination of the parameters of these oscillations and their systematic measurement control in the operating mode are of interest for the assessment of the technical condition of the GTE and its resource.

Torsional vibrations are one of the factors affecting the life of the GTE shafts. Determination of torsional vibration parameters is necessary due to the fact that they are one of the main causes of defects in the rotors. They are also an important parameter characterizing the shaft rotation mode. For practical registration of torsional oscillations is proposed chronometric control.

Chronometric control of the shaft rotation mode of the HD provides high-precision measurement of the duration of its rotation of the shaft and the proportion of turnover . Mathematical processing of the obtained time series allows to calculate the spectral characteristics of the motion. However, the direct space-time structure of torsional oscillations is also informative. Below is a technique for obtaining the parameters of low-frequency torsional vibrations of the shaft based on precision chronometric measurements.

Let the angle of rotation of the rotating shaft of the job expression:

$$\Psi(t) = \omega_{н} t + \phi(t), \tag{1}$$

where $\omega_{н}$ — rated angular velocity, $\phi(t)$ — the perturbation with respect to uniform rotation..

Along with systematic measurements of each period, n fractions are measured. Thus, n is the number of pulses per revolution. From the totality of all time counts $\{t_i^k\}$, где where i — is the period number, $k\ (1 \le k \le n)$ — the number of the count within each period, we can distinguish n subsequences of periods of the elements $\{T_i^k\}$ which are shifted at intervals of Δt_i^k. The number of intervals is the result of the measurement. The number of intervals $\Delta t_i^k = \frac{1}{n} T_i$ is the result of the measurement.

For each turn right, that:

$$\begin{gathered} 2\pi = \int_{t_{i-1}}^{t_i} \dot{\Psi}(t)dt = \omega_{н}^k \left(\delta t_i^k - \delta t_{i-1}^k\right) + \\ +\phi\left(iT_{н}^k + \delta t_i^k\right) - \phi\left[(i-1)T_{н}^k + \delta t_{i-1}^k\right]. \end{gathered} \tag{2}$$

When $\delta t \ll T_i^k$, the ratio is given:

$$\begin{gathered} \left[\omega_{н}^k + \dot{\phi}((i-1)T_{н}^k)\right]\delta t_{i-1}^k + \phi((i-1)T_{н}^k) = \\ = \left[\omega_{н}^k + \dot{\phi}(iT_{н}^k)\right]\delta t_i^k + \phi(iT_{н}^k), \end{gathered} \tag{3}$$

when $1<i<N$.

In the system of equations of magnitude $T_{н}^k, \omega_{н}^k, \delta t_i^k$ are known, $\phi(iT_{н}^k)$ — you must to define

$$\left[\omega_{н}^{k_0} + \dot{\phi}(iT_{н}^{k_0})\right]\delta t_{i_0}^{k_0} + \phi(iT_{н}^{k_0}) = 0. \tag{4}$$

If $i = i_0, k = k_0$ corresponds to the groove number **k** can be taken as the starting point

If long periods of perturbation $\dot{\phi}(iT_{н}^k) \ll \omega_H^k$, then

IOP Conf. Series: Materials Science and Engineering **714** (2020) 012010 doi:10.1088/1757-899X/714/1/012010

$$\phi\left(iT_{н}^{k}\right) \cong -\omega_{н}^{k} \cdot \delta t_{i}^{k}. \quad (5)$$

and the chronogram is converted into the law of motion of the shaft with a sampling step in time equal to the nominal period of rotationIf additional measurement information is required for processing real chronograms.

Instead $\delta t_{i_0}^{k_0} = 0$ of equality it is enough to demand $\left|\delta t_{i_0}^{k_0}\right| \le \Delta t$. where Δt — absolute error of measurement of time intervals. This approach makes it possible to present the law of motion of the elements of the shaft line in polar coordinates and to control the process of its twisting using chronometric information in different sections.

3. Modern diagnostics of gas turbine engines by phase-chronometric method

Most decisions in science and technology are made on the basis of measurement information and its processing. Information and metrological support of the life cycle of machines and mechanisms includes methods and means of obtaining measuring information and its processing for the assessment of the current technical condition of cyclic machines and mechanisms; emergency protection, research tasks of technical objects, management.

As an example of implementation, the application of phase-chronometric method of control and diagnostics of gas turbine engine (GTE) is considered.

The mathematical model of the rotor consists of a system of differential equations describing the bending-torsional vibrations of the GTE rotor [18].

The following assumptions are introduced in the calculation of the dynamic scheme of the gas turbine engine:

- the multi-disc rotor can be replaced by an equivalent system with a much smaller number of disks by combining them;
- shafts and housings are presented in the form of systems with distributed mass or in the form of a chain of discrete systems;
- thin-walled structural elements are schematized in the form of shells, plates, rings;
- the details connecting the bearings with the outer hulls, are considered to be elastic, their masses to join the masses of the case and of the bearings may be tailored to give the bearings and supports;
- when the console disc attachment to the shaft takes into account the yielding of the disks under their cloths deformation;
- the shoulder blades are considered to be absolutely rigid.

The front support of the fan rotor is elastic and is made with a ball angular contact bearing. Back support-rigid, made with roller bearing.

The operation of a gas turbine engine is characterized by turbine moments and compressor resistance.

The transient operation mode is characterized by a section at which the moment changes dramatically with this mode of operation, the mechanical running parts of the GTE are most prone to wear.

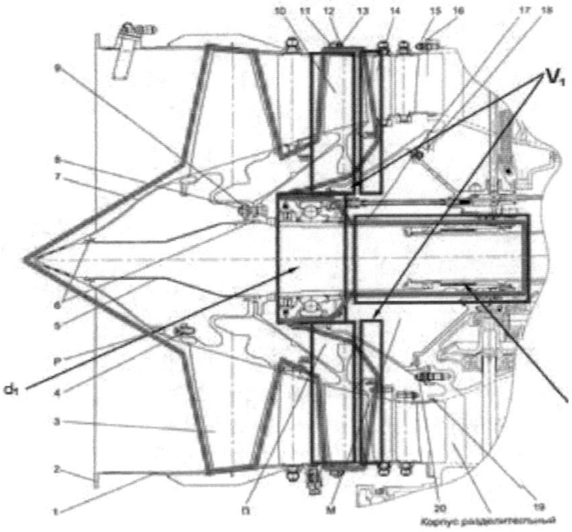

Figure 1. Fan with rotor, divided into count areas.

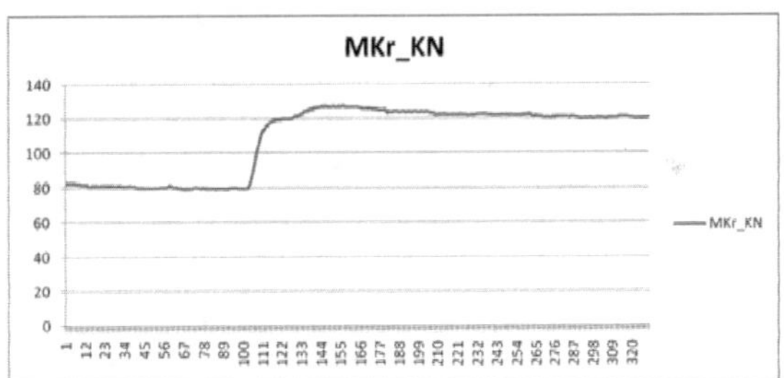

Figure 2. The torque on the low-pressure compressor.

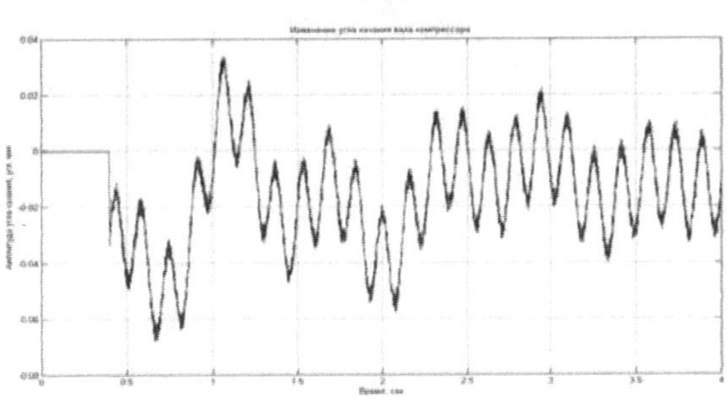

Figure 3. Changing the swing angle of the compressor shaft, angle./mine.

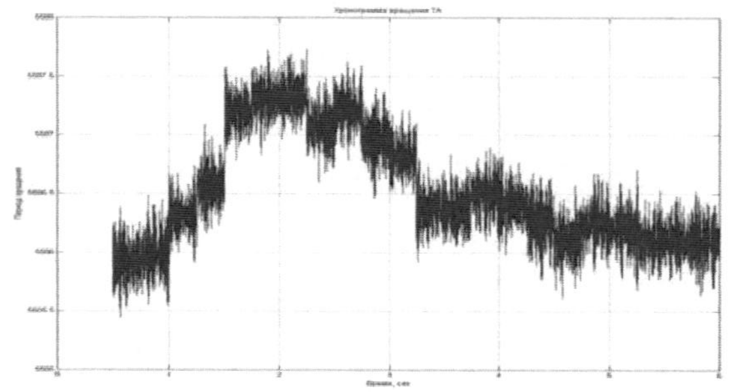

Figure 4. Simulated chronogram of rotation at steady-state operation, n = 10750 vol./mine.

The results of the computational experiment allow us to draw the following conclusions:

- changing torsional stiffness affects the value of the natural frequencies of torsional vibrations;
- the application of the refined multi-factor mathematical models of objects in a phase-chronometric representation allows to obtain a more detailed assessment of the functioning of the gas turbine simulation of critical conditions and correlation of measurement results with the design of the GTE.

The results of mathematical modeling of measurement control that have been presented here reveal essentially new capabilities for the measurement and estimation of the degradation of the properties of the structural materials of the shaft lines of turbine units in the course of service, including:

1. a solution of the problem of detection of cracks at an early stage of their development through measurement of the deviation of the frequency of the torsional vibrations of the shaft line;

2. determination of external cyclic loads on the shaft line and the moving parts of mechanisms in the course of service;

3. an increase in the precision of the determination of the physical parameters and properties of structural materials by several orders of magnitude and, as a consequence, a reduction in the time and an increase in the reliability of technological-design development.

The results that have been obtained may be used in basic studies of the evolution and degradation of the physico-mechanical properties of structural materials.

4. Conclusion

It seems relevant to use the possibility of using modern tools and methods of precision phase chronometry to obtain information about the functioning of gas turbine engines.

On the basis of precision chronograms, obtaining of spatiotemporal characteristics of low-frequency torsional vibrations of the shaft rotation is realized, where the role of the carrier frequency modulated by the low-frequency signal is played by the nominal speed of the shaft. Thus, there is an additional possibility of systematic monitoring of the state of gas turbine engines.

References

[1] Korsun O N, Nikolaev S V and Pushkov S G 2016 *Journal of Computer and Systems Sciences International* **3** 446–57
[2] Giusti A and Mastorakos E 2019 *Flow Turbulence Combust* https://doi.org/10.1007/s10494-019-00072-6
[3] Geigle K P, O'Loughlin W, Hadef R and Meier W 2015 *Appl. Phys. B* **119** 717–30
[4] Shultz C and Sick V 2005 *Prog. Energy Combust. Sci.* **31** 75–121
[5] Magnotti G and Barlow R 2017 *Proc. Combust. Inst.* **3** 4477–85
[6] Schildt P, Braun C and Marzocca P 2019 *CEAS Aeronaut J* https://doi.org/10.1007/s13272-019-00411-3
[7] Tinga T and Loendersloot R 2019 Physical Model-Based Prognostics and Health Monitoring to Enable Predictive Maintenance *Predictive Maintenance in Dynamic Systems* Lughofer E, Sayed-Mouchaweh M (eds) (Cham: Springer)
[8] Tran V T, Thobiani F A, Tinga T, Ball A D and Niu G 2017 *Proc. IME C. J. Mech. Eng. Sci.* **232** 3767–80 https://doi.org/10.1177/0954406217740929
[9] Breteler D, Kaidis C, Tinga T and Loendersloot R 2015 Physics based methodology for wind turbine failure detection, diagnostics & prognostics *EWEA Rosmi A. (ed.)* p 9
[10] Thobiani F A, Tran V T and Tinga T 2017 *Engineering* **6** 524–39
[11] Hu C, Youn B D and Wang P 2019 Time-Dependent Reliability Analysis in Operation: Prognostics and Health Management *Engineering Design under Uncertainty and Health Prognostics. Springer Series in Reliability Engineering* (Cham: Springer)
[12] Sun J, Zuo H, Wang W and Pecht MG 2012 *Mech Syst Signal Process* **28** 585–96
[13] Xu J, Wang Y and Xu L 2014 *Sensors J* **4** 1124–32
[14] Wang T, Yu J, Siegel D and Lee J 2008 A similarity-based prognostics approach for remaining useful life estimation of engineered systems *2008 International Conf. on Prognostics and Health Management* doi: 10.1109/PHM.2008.4711421
[15] Zhang S and Ganesan R 1997 *J Eng Gas Turbines Power* **2** 378–84
[16] Varaksin A Y and Orlov M A 2018 *Doklady Physics* **9** 385–7
[17] Mochalov A A, Varaksin A Y and Arbekov A N 2019 *Measurement Techniques* **3** 242–8
[18] Pronyakin V I 2017 *MATEC Web of Conferences* **99** 02006
[19] Metelkina E and Komshin A 2017 *MATEC Web Conf.* **129** 01048
[20] Kiselev M I, Komshin A S, Pronyakin V I, Korsun O N and Nemichev M Yu 2011 *Measurement Techniques* **54** 1081–90

Comparative Study of Pontryagin Maximum Principle and Direct Method in Aircraft Optimal Control Design

O N Korsun[1], A V Stulovskii[1] and S P Gulevich[2]

[1] State Research Institute of Aviation Systems, Moscow, Russia
[2] VR-Technologies, Moscow, Russia

E-mail: marmotto@rambler.ru

Received xxxxxx
Accepted for publication xxxxxx
Published xxxxxx

Abstract

This paper deals with the problem of open-loop control design for a dynamic object which is described by a nonlinear system of differential equations. To solve this problem a direct method is used, in which the target functional is minimized by population-based algorithm. The method is applied to a test problem which consists of finding optimal control for spacial motion of maneuverable aircraft. Proposed technique is compared with two classical solutions to considered problem. One of them is based on equating to zero the partial control derivatives of the Hamilton function, wherease the other on the Hamilton function maximum over controls (Pontryagin maximum principle). The paper confirms high degree of similarity between solutions obtained by all considered methods of selecting target functional. However, classical algotithms show slightly worse accuracy and higher sensitivity to the quality of initial approximation. In addition, proposed direct method permits to evade the necessity to solve a two–point boundary value problem required in classical algorithms.

Keywords: optimal open-loop control, Pontryagin maximum principle, population-based numerical optimization algorithm

1. Introduction

A number of ways exist for solving the open-loop control problem [1], starting from the classical results of the calculus of variations and the Pontryagin maximum principle and continuing with relatively new developments such as methods of inverse dynamic problems [2], decomposition [3] and embedded systems.

In recent decades considerable attention have been attracted to the direct methods [4, 5]. These techniques are based on the assumption that possible control could be parameterized and, thus, reduce the control search to a single-criterion unconditional multi-parameter optimization problem. In such formulation the problem appears much simpler than traditional two-point boundary value problem. Since the use of many well-proven gradient methods (including the Newton method with its modifications) is difficult for a large number of parameters, turn to another type of optimization algorithms, namely, genetic and population-based ones looks reasonable. These algorithms have shown encouraging results for multidimensional optimization problems, therefore we use the particle swarm method [6, 7], belonging to the population-based algorithms.

The paper proposes a direct method for the formation of open-loop control of an aircraft and compares it with the classical results of optimal control theory — solution to the Lagrange problem obtained on the basis of the calculus of variations and the Pontryagin maximum principle.

IOP Conf. Series: Materials Science and Engineering **714** (2020) 012011 doi:10.1088/1757-899X/714/1/012011

2. Specification of problem

Let us assume that nonlinear dynamical object model takes the form

$$\dot{\mathbf{x}} = \mathbf{f}(\mathbf{x}, \mathbf{u}, t), \quad (1)$$

where $\mathbf{x}$ and $\dot{\mathbf{x}}$ — state or output vector and its time derivatives, $\mathbf{u}$ — vector of control or input vector, $\mathbf{f}(\cdot)$ — vector function of vector argument. Initial conditions $\mathbf{x}(t_0) = \mathbf{x_0}$ are considered known.

Targeted functional

$$J(\mathbf{x}, \mathbf{u}) = \int_{t_0}^{T} F(\mathbf{x}, \mathbf{u}, t)\, dt \quad (2)$$

is determined on time interval $t \in [t_0; T]$.

The task is to find such control that provides for the considered time interval the minimum deviations of the output signals from the given ones. Then in target functional integrand can be written as follows

$$F(\mathbf{x}, \mathbf{u}, t) = F(\mathbf{x} - \tilde{\mathbf{x}}, \mathbf{u}, t), \quad (3)$$

where $\tilde{\mathbf{x}}$ — vector of given output signals.

Expressions (1)–(2) describe the well-known Lagrange problem with a free end point, since the initial conditions at the left end are given and the boundaries of the time interval are fixed. The traditional recommendation to find the optimal control is the transition to a two-point boundary value problem. To do this, the Hamilton function is written

$$H = (\boldsymbol{\lambda}, \mathbf{f}) - F,$$

where conjugate vector function is obtained from equation

$$\dot{\boldsymbol{\lambda}} = -\mathbf{f}_y^T \boldsymbol{\lambda} + F_y, \quad \boldsymbol{\lambda}(T) = 0, \quad (4)$$

Optimal control is found from necessary optimality condition

$$\frac{\partial H}{\partial \tilde{\mathbf{u}}} = \boldsymbol{\lambda} \mathbf{f}_u - F_u = 0, \quad (5)$$

$\tilde{\mathbf{u}}$ here means optimal control.

The two-point boundary-value problem arises, since the initial conditions for the object model (1) are given at the beginning of the interval, and for the conjugate function (4) at its end. Search for solution to this problem is often fraught with significant difficulties. The form of condition (5) corresponds to the case when restrictions are not imposed on the control.

If there are restrictions, the necessary minimum condition is determined by the Pontryagin maximum principle

$$H(\tilde{\mathbf{x}}, \tilde{\mathbf{u}}, \psi, \tau) = \max_{\mathbf{u} \in G_u} H(\tilde{\mathbf{x}}, \mathbf{u}, \psi, \tau), \quad (6)$$

where $\tilde{\mathbf{u}}$ is chosen from set of feasible controls $\mathbf{u} \in G_{\mathbf{u}}$.

Let us now consider the proposed direct method. It consists of the minimization of the target functional in the form (3).

To do this the control signals, which are functions of time, are represented on the considered interval in the form of a cubic Hermite spline. The spline parameters are a priori unknown. To find them, the problem of multidimensional parametric optimization is solved, i.e. the parameters that deliver the minimum to the functional (3) are found. We find a numerical solution using one of the population optimization methods — the particle swarm algorithm.

Functional (3) in its physical sense reaches a minimum when the model output signals corresponding to optimal control are closest to the given output signals. Therefore, the theoretical basis for this approach is the well-known theorem of the existence and uniqueness of the solution for a finite-dimensional system of nonlinear differential equations. From a practical point of view, the advantage of the direct method is that it does not require the solution of a two-point boundary value problem.

3. Control object

The model of the object takes the form presented below

$$\begin{aligned}
\dot{\alpha} &= \omega_z - \frac{qS}{mV} c_{ye}(\alpha) - \frac{P_x}{mV}\sin(\alpha) + \\
&+ \frac{g}{V}(\sin(\alpha)\sin(\nu) + \cos(\alpha)\cos(\nu)\cos(\gamma)), \\
\dot{V} &= -\frac{qS}{m} c_{xe}(\alpha) + \frac{P_x}{m}\cos(\alpha) + \\
&+ g(-\cos(\alpha)\sin(\nu) + \sin(\alpha)\cos(\nu)\cos(\gamma)), \\
\dot{h} &= V(\cos(\alpha)\sin(\nu) - \sin(\alpha)\cos(\nu)\cos(\gamma)), \\
\dot{\psi} &= \frac{1}{\cos(\nu)} \omega_z \sin(\gamma),
\end{aligned} \quad (7)$$

where α, β — angles of attack and sideslip, rad; ω_z — angular velocity with respect to the axis Oz, rad/s; ν, γ, ψ — angles of pitch, roll and yaw, rad; V — airspeed, m/s; h — flight altitude, m; $c_{xe}(\alpha)$ — drag coefficient; $c_{ye}(\alpha)$ — lift coefficient; m — aircraft mass, kg; S — wing surface area, m^2; P_x — projection of thrust force on axis Ox of body-fixed coordinate system, N; g — gravity acceleration, m/s^2; q — dynamic air pressure, Pa.

The system of differential equations is also supplemented by an algebraic equation for ω_z

$$\omega_z = \dot{\nu}\cos(\gamma) - \dot{\tilde{\psi}}\cos(\nu)\sin(\gamma), \quad (8)$$

where $\tilde{\psi}$ — required value of yaw angle.

For the given model, we choose pitch and roll angles as control signals. They have the advantage that, firstly, they are quite smooth and therefore can be well approximated by splines with a small number of nodes, and secondly, by definition, they have restrictions on the range of values (±90° in pitch angle, ±180° in roll angle). The practical implementation of these angles also does not cause fundamental difficulties, since the methods for designing tracking contours for given pitch and roll angles are currently developed quite well.

Direct methods for solving the problem require expansion of control signals in the basis of linearly independent functions. In this case, it is necessary to strive for a compromise between the accuracy of approximation of the signal, which leads to an increase in the number of parameters, and the computational complexity of solving the parametric optimization problem. In this work, we used Hermite splines of the third order [8] with a distance between nodes of about 8 s.

To test the proposed method, a specially formulated problem was solved. Known signals were set as an input. They were used to model the movement of the object according to the system (7, 8). Resulted output signals were taken as given output $\tilde{\mathbf{x}}$. It was further assumed that the model of the object is known, and the task was set to search for control that minimizes target functional

$$J(\mathbf{x},\mathbf{u})=\sum_{i=0}^{N}(k_1(\alpha(t_i)-\tilde{\alpha}(t_i))^2+k_2(V(t_i)-\tilde{V}(t_i))^2)+$$
$$+\sum_{i=0}^{N}(k_3(h(t_i)-\tilde{h}(t_i))^2+k_4(\psi(t_i)-\tilde{\psi}(t_i))^2), \qquad (9)$$

where k_1,k_2,k_3,k_4 — weight coefficients, $\tilde{\alpha},\tilde{V},\tilde{h},\tilde{\psi}$ — given values of angle of attack, airspeed, flight altitude and angle of yaw.

Target functional for Lagrange problem was formulated differently. Provided that condition of equation of control derivatives to zero is hardly obtainable for numerical calculation, proximity measure of control derivatives to zero was established on considered interval

$$J(\mathbf{x},\mathbf{u})=\sum_{i=0}^{N}\left(k_1\left|\frac{\partial H}{\partial u_1}\right|_{t_i}^2+k_2\left|\frac{\partial H}{\partial u_2}\right|_{t_i}^2\right). \qquad (10)$$

4. Results

Main results concerning operation of direct method and the classic two-point boundary value problem solution in test examples where control signals do not come close to their limit values are presented in [9]. This corresponds to the Lagrange problem (1, 2). If there are restrictions on control, the classical theory recommends applying the maximum principle, where the target functional is determined by expression (6).

So, using a problem with restrictions as example, we compare three methods: the direct method, the solutions based on the equality of the derivatives of the Hamilton function to zero, and the solution by the maximum principle, i.e., by the maximum condition of the Hamilton function (6). Examples are chosen in such way that the desired control signals reach their limits. We introduce a limit of ±40° in pitch angle and ±90° in roll angle.

We conduct the numerical experiment described in Section 3, taking into account the fact that the control signals cannot go beyond the above limits. As can be seen in figure 1, the application of the direct method with functional (9) allows us to achieve the high degree of correspondence in input signals.

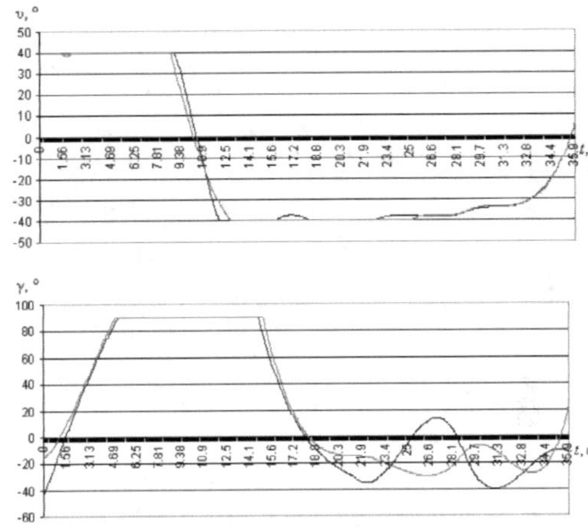

Figure 1. Comparison of given input signals (blue line) with input signals obtained using the direct method (violet line).

Using the functional of the derivatives of the Hamilton control function (10) (figure 2) leads to results close to those obtained previously (figure 1). The accuracy of the estimates, as in the case without limitations, is quite high, although somewhat inferior to the direct method.

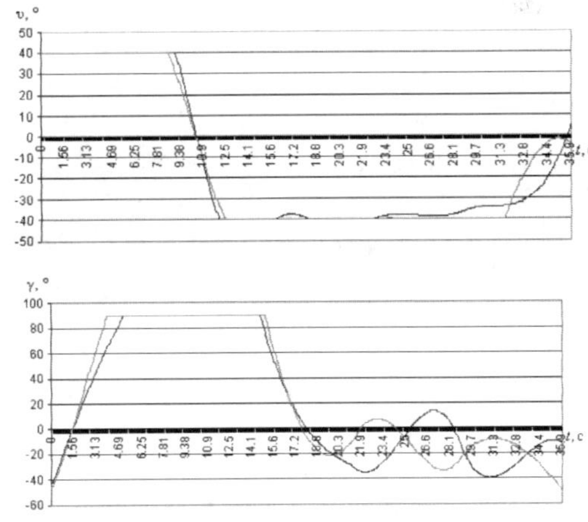

Figure 2. Comparison of given input signals (blue line) with input signals obtained using the partial control derivatives of the Hamilton function (violet line), functional (10).

To numerically find the maximum of the Hamilton function, condition (6) must be specified. We propose the following approach. Suppose that at some stage of the operation of the numerical optimization algorithm M control options are obtained, each of them has its own Hamilton function $H_i(t_j), i=\overline{1,M}$, calculated for N discrete time instants $t_j, j=\overline{1,N}$. Among these functions, we consider the maximum one that at any moment of time t_j, or at most of such moments, surpasses all others. Consider the moment of discrete time t_j for which M values of the Hamilton function are determined. We assign a rank r_j^i to each of these values according to the following rule: the maximum value corresponds to a rank of 1, the next to a rank of 2, and so on. Then each function $H_i(t_j)$ and, accordingly, each control $\mathbf{u}_i$ can be associated with a generalized rank, calculated by the formula

$$J(\mathbf{u}_i)=(\sum_{j=1}^{N} r_j^i - N)^2. \qquad (11)$$

Let us explain the meaning of the generalized rank (11). Suppose that for some control the Hamiltonian function is the largest at all time moments, that is, at each instant it has rank 1. Then the generalized rank (11) is equal to zero and has the minimum possible value. This allows us to reduce the problem of finding the maximum Hamilton function to the problem of minimizing functional (11).

The results of processing the test example, obtained when parameter values are ±20% of actual value, are presented in figure 3.

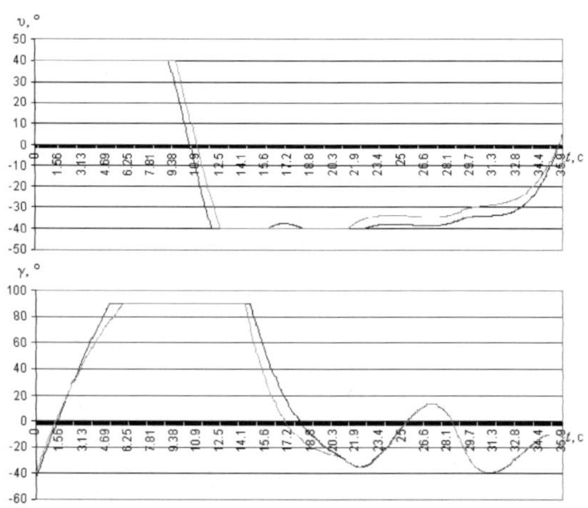

Figure 3. Comparison of given input signals (blue line) with input signals obtained using an algorithm based on Hamilton function maximum (violet line), functional (11).

Note that for a wide search area (±90° in pitch, ±180° in roll), the convergence of the algorithm to the true values is not guaranteed. Figure 3 shows that the obtained accuracy with respect to the input signals approximately corresponds to the above-considered results of the direct method (figure 1) and the method based on the derivatives of the Hamilton function (figure 2), although it is somewhat inferior to them.

5. Conclusions

A direct method for finding the optimal open-loop control based on the parameterization of the control signal and the direct minimization of the functional using the population algorithm is proposed. The efficiency of the method is confirmed by mathematical modeling.

The paper also compares the direct method with two classical approaches based on equating to zero the partial control derivatives of the Hamilton function and the maximum of the Hamiltonian control function (maximum principle). Comparison is performed using test problem describing the spatial motion of a maneuverable aircraft. In the framework of the considered example, all methods showed similar results. At the same time, the accuracy of classical algorithms appears to be slightly worse, and they showed a higher sensitivity to the quality of the initial approximation.

Thus, the practical significance of the results obtained is that, in comparison with classical algorithms, the application of the direct method is significantly simpler, at least for the considered class of control problems, when the controls are sufficiently smooth functions and the right-hand sides of the differential equations of the object are continuous and have continuous first derivatives. In the direct method, for example, cumbersome mathematical calculations necessary for obtaining the conjugate vector function are not required.

Acknowledgements

This work was supported by Russian Foundation for Basic Research (RFBR), project 18-08-00921-a.

References

[1] Pupkov K A and Egupov N D. ed 2004 *Methods of classical and modern theory of automatic control* (Moscow: Izd. MGTU im. Baumana)
[2] Krut'ko P D 1988 *Inverse problems of the control systems dynamics. Nonlinear models* (Moscow: Nauka)
[3] Chernous'ko F L 1995 *Journal "Trudy matematicheskogo instituta im. V.A. Steklova RAN"* **211** 457–72
[4] Hargraves C R and Paris S W 1987 *Journal of guidance, control, and dynamics* **10** 338–42
[5] Von Stryk O and Bulirsch R 1992 *Annals of operation research* **37** 357–73
[6] Olsson A E 2011 *Particle swarm optimization: theory, technique and applications* (Hauppage, USA: Nova Science Publishers)

[7] Mirjalili S, Song Dong J and Lewis A ed 2019 *Nature-inspired optimizers: theories, literature reviews and applications* (Springer)

[8] Zav'yalov Ju S, Kvasov B I and Miroshnichenko V L 1980 *Methods of spline functions* (Moscow: Nauka)

[9] Korsun O N and Stulovskii A V 2019 *Journal of Computer and Systems Sciences International* **2** 229–43

IOP Conf. Series: Materials Science and Engineering **714** (2020) 012012 doi:10.1088/1757-899X/714/1/012012

Operator's State Estimation Based on the Face's Video Images Analysis Using Deep Convolutional Neural Networks

O N Korsun[1], V N Yurko[1] and E I Mikhaylov[1]

[1]State Research Institute of Aviation Systems, Moscow, Russia

E-mail: marmotto@rambler.ru

Received xxxxxx
Accepted for publication xxxxxx
Published xxxxxx

Abstract

The paper deals with the problem of operator's state estimating. For this purpose various approaches based on using deep convolutional neural networks are proposed. The approach using automatic emotion recognition methods is considered in the most detail. During the experiment video records of the operator's face registered during operator performing the flight task on the flight simulator were processed. To determine the type of operator's activity the studies based on using the emotional background of the face are also carried out. The experimental results of this approach confirmed the efficiency of the selected methods, especially for monitoring the operator's state when falling asleep.

Keywords: convolutional neural network, emotion, operator's state

1. Introduction

During the analysis of the operator's face video images using deep convolutional neural networks two approaches were applied:

- neural network was trained according to the traditional recognition way in order to distinguish normal operator's state and fatigue. For this training experimental data received during operators perform flight task was used;
- possibility of using the emotions automatic recognition by face video images was investigated. Now there are many databases from the open access that can be used for neural network training.

Nowadays deep convolutional neural networks are considered the most perspective approach to pattern recognition tasks. The peculiarity of the deep learning methods is that it gradually, layer by layer constructs more and more complex representations and takes into account their interaction, so each layer is updated in accordance with the needs of both the previous and subsequent layers. There are very functional software tools for research in the field of deep learning that effectively implement the back propagation method and operations with tensors, which greatly simplifies the realization of new structures of neural networks.

Emotion is a special kind of medium duration psychic processes, representing a subjective evaluative attitude to existing or possible situations and the objective world. Despite the individual and cultural differences between people, there are general, genetically determined paradigms that determined exactly how our emotions are expressed as a sort of contractions of perfectly defined groups of facial muscles: forehead, eyebrows, eyelids, cheeks, lips, chin.

2. Description of the research and results of the experiments

The following are the stages of automatic emotions recognition:

1. Real-time image registration by capturing the video stream of the camera, or previously recorded video files analysis;

2. Recognition of facial contours in the image;

3. Facial images transmit to a trained convolutional neural network and the emotions classification results output.

For the image facial contours recognition, the Haar cascade was chosen. This is one of the methods for recognizing objects' classes with a high speed of operation. In this article an approach based on the sliding window technology was used: the image was scanned by the sliding window and the Haar attribute was calculated for each image area scanned by the window. The object presence or absence in the window was determined by the difference between the value of the attribute and the threshold value, that was being trained. This system is completely automated and does not require human intervention, so this approach works quickly.

For deep convolutional neural network training the FER-2013 data set [1] was chosen. This database consists of 35887 monochrome images, 48 × 48 pixels in size, with 7 types of emotions (anger, disgust, fear, joy, sadness, astonishment and neutral state).Herewith there were 28709 images used for training, 3589 images — for checking and 3589 for testing. Data from the database was converted to tensors with real numbers, pixel values were scaled from the range [0 – 255] to the range [0 – 1], cause it is unsafe to transmit to the neural network data that can take very large values or dissimilar data. Failure to these rules can result in significant gradient changes that will prevent network convergence.

The architecture of the convolutional neural network constructed in this work with the use of the accepted terminology is presented in the figure 1.

[2×Convolution (3×3)]— MaxPooling (2×2)— Dropout (0.5),
[2×Convolution (3×3)]— MaxPooling (2×2)— Dropout (0.5),
[2×Convolution (3×3)]— MaxPooling (2×2)— Dropout (0.5),
[2×Convolution (3×3)]— MaxPooling (2×2)— Dropout (0.5),
Dense (512)— Dropout (0.5),
Dense (256)— Dropout (0.5),
Dense (128)— Dropout (0.5),
Dense (7)

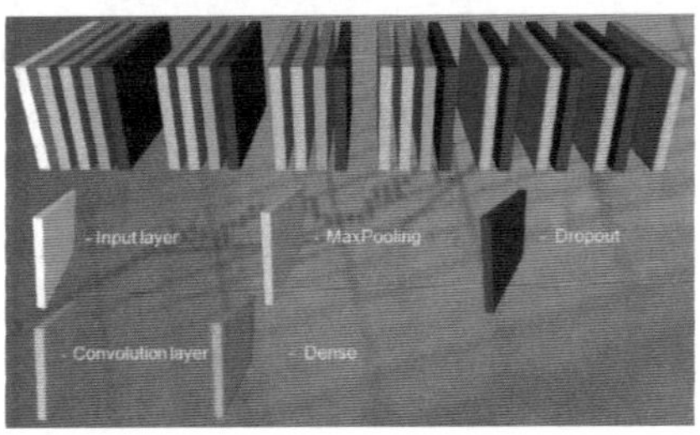

Figure 1. The architecture of the convolutional neural network.

Convolution layers use the ReLU (rectified linear unit) activation function. First convolutional layer contains weights regularization to reduce the network complexity by limiting its weight coefficients. In this article L2-regularization was used whose main idea is that the penalty added to the loss function is proportional to the squares of the weight coefficients. Also in the work "dropout" was used as one of the most effective and common regularization methods for neural networks. Dropout consists in the randomly selected attributes removal at the training stage (the percentage of attributes is chosen in accordance with the dropout coefficient).The last completely connected layer is the loss layer, which returns an array with seven probability estimates.

Examination of the proposed structure on a test set containing 3589 images showed 62.4% of correct recognitions. The received results analysis shows that it is possible to improve the neural network architecture in order to reach the level of 92% correct recognitions.

Further on the trained neural network there was conducted an experiment to estimate the operator's state. During the experiment 7 video records of the operator's face received during operator performing the flight task on the flight simulator were processed. The flight task durations were 14, 18, 19, 20, 12, 19, 12 minutes. The frame rate in the video records was 24 Hz.

In the normal operator's state during the piloting there were 2 emotions observed – sadness and a neutral state (figure 2).It should be noted that the operator's yawn and closed eyes were classified as astonishing or fear (figure 3).

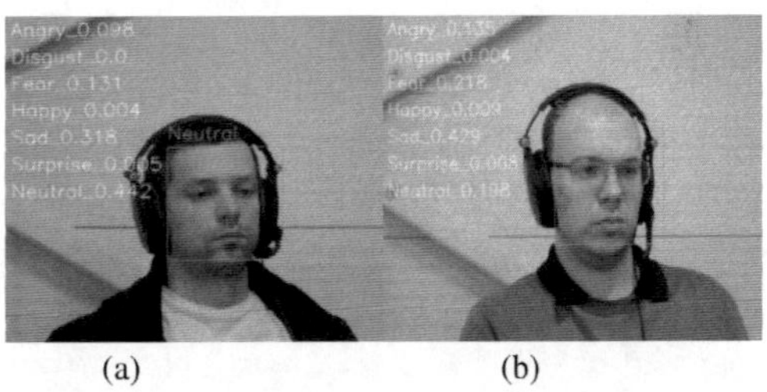

(a) (b)

Figure 2. Neutral state (a) and sadness (b).

(a) (b)

Figure 3. Yawn classification as astonishment (a) or fear (b).

Table 1. The results of emotions classification into the categories after processing seven flight records using a convolutional neural network (the recognized emotions percentage of the total emotions' number during the performing a flight task is presented).

	Anger	Disgust	Fear	Joy	Sadness	Astonish	Neutral
Pilot1	2.4	0	2.1	0	36	4	56
Pilot2	0	0	0	0	37.7	0	62.3
Pilot3	21	0	0.2	0	78.3	0	0
Pilot4	0	0	0.06	0	1.8	1.45	96.7
Pilot5	0.1	0	1.37	0	6.7	3.88	87.9
Pilot6	0.02	0	0.78	0	1.8	1.62	95.75
Pilot7	0	0	1.53	0	2.21	8.05	88.2

The obtained results have an obvious physical interpretation. Operators perform monotonous work during a long time, which corresponds to a normal or sad facial expression. The tendency to fall asleep, expressed in yawning, is classified as astonishment or fear, because yawning as a separate class was not introduced during the neural network training.

According to psychologists [2] fatigue like a fear causes the desire to stop doing something and, if activities do not change, the operator's body will go into a state of sleep.

The following experiments were carried out to study operator's fatigue. A deep convolutional neural network with a trained convolutional basis of the VGG16 [3] model was constructed (it was trained for 1.4 million images belonging to 1000 classes); the last convolutional layer and a fully connected classifier were retrained. The output layer included two neurons corresponding to the operator's states: normal or fatigue (tired). Training database consisted of images data sets from the first and last minutes of the operator's face record during the flight task performance.

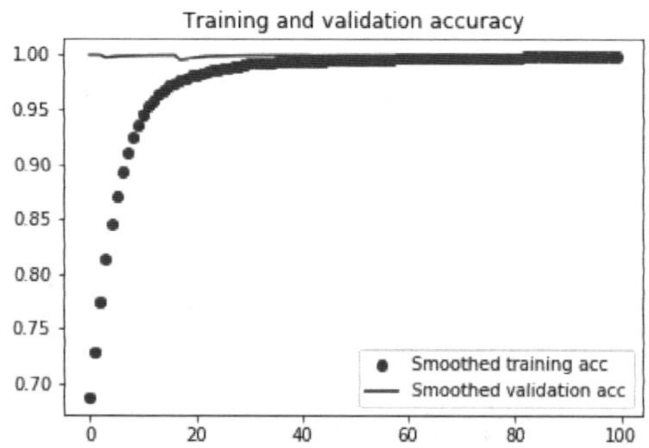

Figure 4. Smoothed accuracy curves at the testing and training stages.

In table 2 the recognized emotions percentage of the total emotions' number during the flight record analysis (record duration was 14 minutes) is presented.

Table 2. The experimental results of distinguishing between two classes ("normal state" and "fatigue").

t, min	Normal	Fatigue
2..3	100	0
3..4	100	0
4..5	100	0
5..6	100	0
6..7	100	0
7..8	95	5
8..9	93.6	6.4
9..10	76	24
10..11	0	100
11..12	5	95
12..13	3	97

Analysis of the record with duration of 14 minutes showed that up to 9.5 minutes the neural network classified the images to the normal state, and then to the "fatigue" state, that confirms the efficiency of the method. Figure 5 shows the image areas by which neural network classifies the video data as a one or another class [4, 5].

Figure 5. Visualization of class activation zones (normal state on the left picture, "fatigue" state on the right picture).

Acknowledgements

This work was supported by Russian Foundation for Basic Research (RFBR), project 18-08-01142-a.

References

[1] Challenges in representation learning: Facial expression recognition challenge http://www.kaggle.com/c/challenges-in-representation-learning-facial-expression-recognition-challenge
[2] Ekman P and Friesen W V 2003 *Unmasking the Face: A Guide to Recognizing Emotions From Facial Expressions* (Malor Books, 8/16/03 edition)
[3] Simonyan K and Zisserman A 2015 *Very Deep Convolutional Networks for Large-Scale Image Recognition. InICLR*
[4] Simonyan K, Vedaldi A and A. Zisserman 2013 *Deep inside convolutional networks: Visualising image classification models and saliency maps CoRR* abs/1312.6034
[5] Selvaraju R R, Das A, Vedantam R, Cogswell M, Parikh D and Batra D 2016 *Grad-CAM: Why did you say that? Visual Explanations from Deep Networks via Gradient-based Localization CoRR* abs/1610.0239

A General Principle of Isomorphism: Integration of Regulator and Observer in the Control System

Vladimir S Kulabukhov[1]

[1] «Avionica» Joint Stock Company, Moscow, Russia

E-mail: kws0704@mail.ru

Received xxxxxx
Accepted for publication xxxxxx
Published xxxxxx

Abstract

The problem is considered and the formalized method of integration of the observer and the regulator in the tracking control system of object of rather arbitrary type is offered. The solution of the problem is carried out on the basis of the General principle of isomorphism. Unlike existing methods, the structure of the control loop is not predetermined. The control loop, the regulator and the observer are determined by the structure of the control object and are formed in a uniquely defined way based on the separation theorem.

Keywords: the principle of isomorphism, the regulator, the observer, the control object, feedback, the separation theorem, integrant, the information-control system

1. Introduction

In [1, 2] the General principle of isomorphism is formulated and a new paradigm in the theory of systems is proposed, which allowed basing on this single principle to substantiate and strictly formalize all the basic concepts and properties of systems (simple, complex, large systems, emergence, causality, observability, controllability, checkability, complexability of systems, as well as postulates and principles of feedback, superposition, relativity, duality, etc.).

In work [2] the General principle of isomorphism allowed to deduce formal conditions of integration (Association) of systems in larger formations-the systems called complexes(complex or integrated systems), and to prove the known separation theorem for systems of very General type. The problems of synthesis or integration of systems into such complexes are very relevant [3–11].

In this article, the results of the General theory of integration [2] are concretized in the Appendix to the formally strict separation of the tasks of synthesis of the observer and the regulator and to their subsequent merging (integration) in a single tracking information-control system (ICS).

2. Description of the research

The problem of correct separation of synthesis tasks and subsequent integration of systems is solved. The following systems are considered: $\Sigma_1 = (U, s_R, X)$ — control system, $\Sigma_2 = (X_0, s_N, Z)$ — observation (evaluation) system, $\Sigma = (X_0, s^*, X)$ — controlled object. The designations are interpreted in accordance with [2, 12–16] and will be clear from the context. As an initial premise, we note that according to [2] the map-structure $s^* : X_0 \to X$ must be isomorphic.

It is necessary to define formal conditions of separation of tasks of synthesis of the regulator $s_R : U \to X$ and the observer $s_N : X_0 \to Z$, and then conditions of Association (integration) of systems of surveillance and control in single ICS of object of control. No preconditions are imposed on the structure of the observation-control loop. The formation of the contour according to formalized rules is also considered as one of the tasks of the synthesis of ICS. It is only required that the object Σ satisfy (defined in [2] by the corresponding theorems) the conditions of observability and controllability with accuracy up to the isomorphism s^*.

For a fixed s^* object structure, a commutative diagram of an integrated system combining a surveillance system and a control system is shown in Fig. 1 [2]. The General conditions for the integration of these systems are determined by the theorem proved in [2]. Its wording is given below.

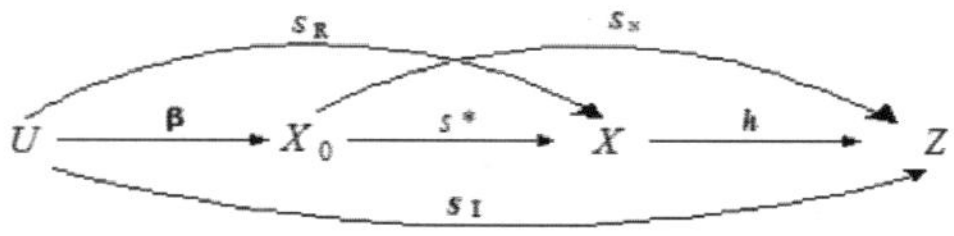

Figure 1. Commutative diagram of an integrated system.

Theorem 1 (theorem of integration). The necessary and sufficient conditions for the integration of the surveillance system $\Sigma_2 = (X_0, s_N, Z)$ and the control system $\Sigma_1 = (U, s_R, X)$ into a single ICS that provides targeted control of the object $\Sigma = (X_0, s^*, X)$ are:

1) the condition of observability $sN = hs^*$ with accuracy up to isomorphism s^*;

2) controllability condition $sR = s^*\beta$ up to the same isomorphism s^*;

3) the General condition of isomorphic determinism of the functioning of the system, determined by the existence of a composition — complex of maps

$$s_I = hs^*\beta = h(s^*\beta) = hs_R = (hs^*)\beta = s_N\beta,$$
$$s_I = hs_R = s_N\beta. \qquad (1)$$

Composition (1) will be called "integrant". In accordance with theorem 1, surveillance and control systems can be combined into an integrated system-complex if and only if there is an integrant (1) that allows formally strictly and unambiguously calculate the vector Z of sensor readings from the control vector U at the input and, conversely, on the known vector Z to calculate the vector U required to obtain it, that is, to close the feedback from Z to U.

Thus, the feedback here is not introduced as a postulate, as is usually done, but is calculated in the form of an integrant (1), the existence of which is a necessary and sufficient condition for the correct unification of the observer and the regulator into a single system. Integrant (1) combines both conditions-the observability condition and the controllability condition of the system with the accuracy of the same model-isomorphism s^*, characterizing the structure of the control object.

The indication "with the accuracy up to model-isomorphism" is very important and means that all statements made are valid strictly within the framework of a commutative diagram built on the s^* isomorphism. The point is that the s^* map must be isomorphic in the usual sense initially-outside of compositions with other maps, outside of the commutative diagram. In this case, it uniquely (with the accuracy up to isomorphism s^*) determines the properties of all other mappings composing with it in the commutative diagram. The meaning of the statement "with the accuracy up to isomorphism" is explained in more detail in [2, 12].

The isomorphism condition of the s^* map is not too strong. For example, all stationary linear dynamical systems, often used in practice, have the property of isomorphism due to the mandatory reversibility of the fundamental matrix of solutions of the corresponding systems of differential equations [5, 13, 14]. In addition, in [2] cases of integration are considered, when the property of isomorphism in the usual sense (before the conclusion in the commutative diagram) has not a map s, characterizing the actual control object, but some other map, for example, h, or β. We will not dwell on these cases here.

According to the realization theorem proved in [2], relying on the General principle of isomorphism, the integrant (1) is an isomorphism with the accuracy up to the isomorphism s^* and is therefore unique in the commutative diagram. This makes it possible to accurately, completely and unambiguously predict the readings of the sensors Z on the vector U and, conversely, the same accurate, complete and unambiguous determination of controls U on the readings of Z sensors. The presence of the isomorphism s^* common to the observer and the regulator provides the possibility of their Union ("gluing") into a single integrated system-complex, that is, ensures the existence of a composition of maps (1), which is an integrant.

The diagram in Fig. 1 is based on the classical definitions of controllability and observability [2]. In practice, according to the indications of Z sensors, it is necessary to restore (observe) not the vector X_0, but the vector X to be regulated at the output of the object. Vector X can be observed using an isomorphic with the accuracy up to the isomorphism s^* observer $\tilde{s}_N$, more precisely, the inverse of it observer

$$(\tilde{s}_N)^{-1} = s^*(s_N)^{-1} = s^*(hs^*)^{-1} = s^*(s^*)^{-1}h^{-1} = h^{-1},$$

where h^{-1} is defined up to the isomorphism s^* by the rules set out in [2, 12–16]. Then the commutative diagram is converted to the equivalent form shown in Fig. 2.

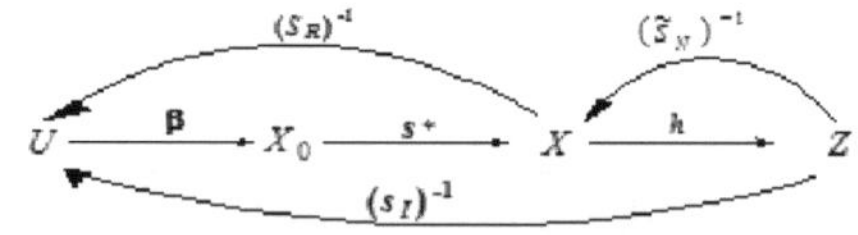

Figure 2. Equivalent commutative diagram of an integrated system.

The diagram in Fig. 2 illustrates the separation theorem for the synthesis of regulator and observer, proved in [2] for the

IOP Conf. Series: Materials Science and Engineering **714** (2020) 012013 doi:10.1088/1757-899X/714/1/012013

General case of arbitrary, but satisfying the conditions of theorem 1, maps h, s^*, β. Its wording is given below.

Theorem 2 (separation theorem). For a dynamical system with an isomorphic structure s^*, the integrant $(s_I)^{-1}$ is a composition of the regulator and the observer

$$(s_I)^{-1} = (s_R)^{-1}(\tilde{s}_N)^{-1}, \quad (2)$$

where

$$(s_R)^{-1} = \beta^{-1}(s^*)^{-1}, \quad (3)$$

$$(\tilde{s}_N)^{-1} = h^{-1}. \quad (4)$$

Theorem 2 follows from theorem 1. If the ICS meets the conditions of theorems 1 and 2, then the necessary and sufficient conditions for the separation of synthesis tasks and the subsequent integration of the observer and the regulator in a single control system are fulfilled.

Separation theorem allows to decompose strictly formally, the General problem of synthesis of a control object into two "independent" tasks: the synthesis problem is of an isomorphic observer, providing the observation of the output X, and the synthesis problem is of an isomorphic regulator that implements feedback from output X to the control inputs U. This provides the ability to control the object in accordance with externally given purpose.

The word "independent" is quoted, because these problems must have a mandatory common element — the same isomorphic model of the object.

First, the problem of observer $(\tilde{s}_N)^{-1}$ synthesis is solved. Then the problem of synthesis of the regulator $(s_R)^{-1}$ is solved. The composition of the feedbacks realized by the observer and the regulator forms a General feedback from Z to U in the form of an integrant (2). The separation theorem allows us to correctly determine the structure of all feedbacks necessary to control the object, which is the practical significance of the separation theorem. In General, the separation theorem proves the necessity and sufficiency of the existence of feedback in the form of integrant (2) for unambiguous, complete and accurate control of the object.

In known works (see, for example, [6–9]) in integration problems, the structure of the observation-control loop is usually postulated in advance. In this regard, it is sometimes concluded that the classical separation theorem is incorrect [5]. In this paper, on the basis of only classical definitions of observability and controllability, the structure of the integrated circuit is determined directly as a result of solving the integration problem, taking into account the isomorphism principle proved in [1, 2]. In addition, the validity of the separation theorem is confirmed in the more General case — for nonlinear systems of General form.

Well-known extreme cases and easily verifiable examples also confirm the correctness of the results obtained. For example, for a scalar object (in which all vectors U, X_0, X, Z have only one component), the regulator in accordance with (3) is determined by the trivial inversion of the scalar "transfer function" of the object W_0 [13, 14], which is the composition $W_0 = s^*\beta$, that is

$$(s_R)^{-1} = (W_0)^{-1} = \beta^{-1}(s^*)^{-1}, \quad (5)$$

and the observer by virtue of the fact that the only component in X in this case is measured directly (that is, $h = 1$) is also trivial and according to (4) is equal to $(\tilde{s}_N)^{-1} = 1$.

For multivariable linear object, which in the General case, the size of the vectors U, X and Z are different, (for example, U — r-vector, X an n-vector and Z an m-vector), the regulator, as before, is determined by the inverse of a "transfer function" of the object W_0 [13, 14], but the formula only in form coinciding with (5). In this case, the inverse β^{-1} and $(W_0)^{-1}$ in (5) must be determined up to the isomorphism s^* according to the rules given in [2, 12]. According to the same rules, the h^{-1} mapping defining the structure of the observer (4) must be defined.

If we introduce the concept of "generalized control object" (GCO), which combines the composition of all three known mappings β, s^* and h, shown in Fig. 2, the control loop is converted to the view shown in Fig. 3. For a linear multi-connected GCO, in whose structure all mappings are written in Laplace operator form, the contour is converted to the form shown in Fig. 4, where $(s_I)^{-1} = \beta^{-1}(s^*)^{-1}h^{-1}$ is a transfer function in operator form. The inverse β^{-1} and h^{-1} are determined up to the isomorphism s^* according to the rules of works [2, 12]. Here the transfer function $W_{GCO} = hs^*\beta$. The outline in Fig. 2 and 3 have a "multiplicative" form characteristic of commutative diagrams characterizing compositions ("products") of mappings [2, 17, 18]. The outline in Fig. 4 has a traditional "additive" form, typical for operations on transfer functions in the operator form [5].

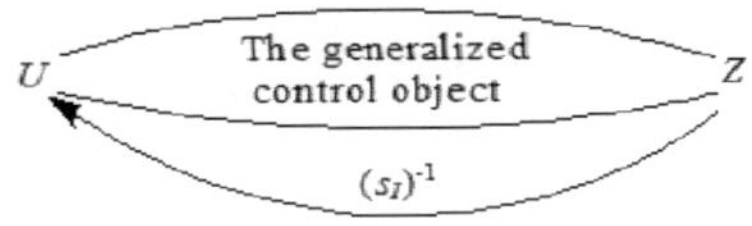

Figure 3. Generalized object control loop.

Figure 4. Control loop of a linear generalized object.

IOP Conf. Series: Materials Science and Engineering **714** (2020) 012013 doi:10.1088/1757-899X/714/1/012013

3. Conclusion

Thus, within the framework of the General principle of isomorphism, based on the classical definitions of controllability and observability, a proof of the separation theorem for the General case of nonlinear systems is given, and a fully formalized method for the synthesis of structures of integrated ICS is proposed.

As follows from the more General formulation and proof of the separation theorem, the behavior of an integrated system is completely deterministic and predictable, that is, an integrated ICS is a simple system in the sense of [2].

References

[1] Kulabukhov V S. 2014 Printsip izomorfnosti v zadache realizatsii i yego prilozheniya k analizu svoystv sistem upravleniya *Proceedings of the XII Vserossiysk, Soveshch po Problemam Upravleniya VSPU-2014, IPU RAN* pp 438–48 http://vspu2014.ipu.ru/prcdngs

[2] Kulabukhov V S 2018 *Cloud of science* **3** 400–72

[3] Kulabukhov V S 2007 Algebraicheskaya formalizaciya procedur analiza i sinteza slozhny`x integrirovanny`x sistem *6 Mezhdunarodnaya konferenciya "Aviaciya i kosmonavtika – 2007"*

[4] Bukov V N, Kulabuxov V S, Kos`yanchuk V V, Ryabchenko V N, Goryunov S V and Naumov A I 2007 *Osnovy` integracii sistem aviacionnogo oborudovaniya. Uchebnoe posobie* (Moscow: Izd. VVIA im. prof. N.E. Zhukovskogo)

[5] Krasovskii A A 1987 *Handbook on the Theory of Automatic Control* (Moscow: Nauka)

[6] Kos`yanchuk V V 2002 *Avtomatika i telemekhanika* **6** 23–35

[7] Kos`yanchuk V V 2004 Sintez integrirovanny`x sistem nablyudeniya i upravleniya s zadanny`mi dinamicheskimi xarakteristikami *Navigaciya i upravlenie dvizheniem. Sbornik dokladov VI konferencii molody`x ucheny`x* 215–21

[8] Kos`yanchuk V V 2005 *Giroskopiya i navigaciya* **48** 100

[9] Kosjanchuk V V, Bukov V N and Ryabchenko V N 2001 *Automation and Remote Control* **3** 15

[10] Kulabukhov V S 2015 *Aviakosmicheskoye priborostroyeniye* **12** 11–31

[11] Kulabukhov V 2019 Algebraic Formalization of System Design Based on a Purposeful Approach *IOP Conference Series: Materials Science and Engineering* **476** 012017 https://iopscience.iop.org/article/10.1088/1757-899X/312/1/012016

[12] Kulabukhov V S 2019 *Symmetry* **11** 1301 doi:10.3390/sym11101301

[13] Kulabukhov V S 2017 *Mechatronics, automation, control* **18** 507–15

[14] Kulabukhov V 2017 Linear Isomorphic Regulators MATEC Web of Conferences, *Proceedings of the CMTAI2016* **9** 03008 doi:10.1051/matecconf/20179903008

[15] Kulabukhov V S 2017 *Radiotekhnika* **8** 50–5

[16] Kulabukhov V S 2017 Isomorphic observers of the linear systems state *IOP Conference Seriesm, Proceedings of the Materials Science and Engineering in Aeronautics (MEA2017)* **312** 012016 http://iopscience.iop.org/issue/1757-899X/312/1

[17] Kostrikin A I 1994 *Vvedenie v Algebru: Osnovy Algebry* (Moscow: Fiziko-Matematicheskaia Literatura)

[18] Bukov V N 2006 *Vlozheniye Sistem. Analiticheskiy Podkhod k Analizu i Sintezu Matrichnykh System* (Kaluga: Izd-vo nauchn. Literatury N.F. Bochkarevoy)

IOP Conf. Series: Materials Science and Engineering **714** (2020) 012014 doi:10.1088/1757-899X/714/1/012014

Automated Identification of Critical Malfunctions of Aircraft Engines Based on Modified Wavelet Transform and Deep Neural Network Clustering

V P Kulagin[1], D A Akimov[1], S A Pavelyev[1] and D A Potapov[1]

[1] Russian Technological University (MIREA), Moscow, Russia

E-mail: akimov_d@mirea.ru

Received xxxxxx
Accepted for publication xxxxxx
Published xxxxxx

Abstract

The paper considers the issues of automatic classification of vibrational states of aircraft engine malfunctions based on the use of convolutional neural network processing of vibrational measurement data presented in spectral form and the knowledge of experts with experience in interpreting spectrograms characterizing the vibrational state of aircraft engines. The developed spectrogram analysis model allows the state monitoring of aircraft engines in automatic mode both during maintenance and in flight operation. The system is able to timely notify technical personnel or crew about the appearance of signs of emergency situations, as well as the type of possible malfunctions. It is shown that the main problem affecting the quality of detection of a potential turbine malfunction is a small sample of data corresponding to malfunctioning states. It is proposed to detect emission anomalies in a small sample by recognizing a modified wavelet transform and neural network clustering, which allows more complete formation of a training sample. The data samples used in training the neural network classifier during the experimental studies were generated on the basis of existing archive files containing complete aperture data from engine vibration sensors and information about malfunctions detected in them.

Keywords: big data, aircraft engines, gas turbine engines, intelligent system, software, deep learning, neural network, recurrent network, vibrodiagnostics, troubleshooting classification, wavelet transform, predicative analytics

1. Introduction

At present in the context of a reduction in the number of personnel, the maintenance staff of aviation equipment remains mainly maintenance specialists, while the number of qualified analysts-experts who are able to analyze spectrograms of data recorded by vibration sensors in continuous monitoring mode is decreasing. Thus, the urgent task is to develop the intelligent systems for monitoring the vibration state of aircraft engines and the detection of signs of malfunctions at an early stage.

In the framework of these studies the approaches have been formulated to provide classifying the vibrational states of gas turbine engines based on the use of neural network processing of data vibrational measurements presented in spectral form and the knowledge of experts with experience in interpreting spectrograms characterizing the vibrational states corresponding to the origin of certain malfunctions.

Gas turbine engines are distinguished by high-speed operation, significant design complexity, and a defect that occurs in one of the structural elements, causing an increase in the level of vibration, has a direct effect on the other elements of the turbine or pump units associated with it [1]. In this

IOP Conf. Series: Materials Science and Engineering **714** (2020) 012014 doi:10.1088/1757-899X/714/1/012014

regard, in recent years, special interest has been shown in the creation of monitoring and vibration diagnostics systems that not only provide for the detection of increased vibration activity of the turbine rotor or pumps, but also, based on the analysis of the obtained vibration information, make it possible to identify the causes of such vibration activity and formulate technical solutions and recommendations for its elimination [2, 3]. Monitoring the machines by the use of such systems is carried out in real time and is necessary to monitor continuously the machine vibration state, in particular, the vibration level of its main components and elements. Diagnosis of defects is carried out on the basis of pre-formed experimental databases and generalized knowledge bases that correlate an increased level of vibration activity with the provided it causes.

2. Turbine failure analysis

Fault analysis by spectrograms of a vibrational signal is one of the most accurate diagnostic methods, and, in connection with this, methods based on it are widespread [4, 5]. The formation of the algorithms set intended for a specific model for obtaining information in the form of diagnostic features is possible with the use of an orthonormal system of basic functions. The feature set is selected on the basis of its acceptability for solving diagnostic problems, the degree of reliability of diagnosis and is also limited by the capabilities of the calculator used — supporting the hardware multiplication, capacity, memory size and other factors. A set of orthonormal feature systems acceptable for spectral diagnostics are limited. The orthogonal trigonometric Fourier functions — the discrete and fast Fourier transform have the widest spreading. Non-trigonometric Fourier series are also common, Chebyshev orthogonal polygons and Walsh functions, in particular.

The tasks of vibro-acoustic diagnostics of gas turbine engines (hereinafter — GTE) can be divided into two groups. The first group of tasks is the determination of the technical condition of the engine and its elements (Figure 1), as well as the early detection of faults to ensure the required engine reliability and reduce costs associated with eliminating the consequences of malfunctions. These problems are solved at the operation stage, during bench tests, during fine-tuning and in production.

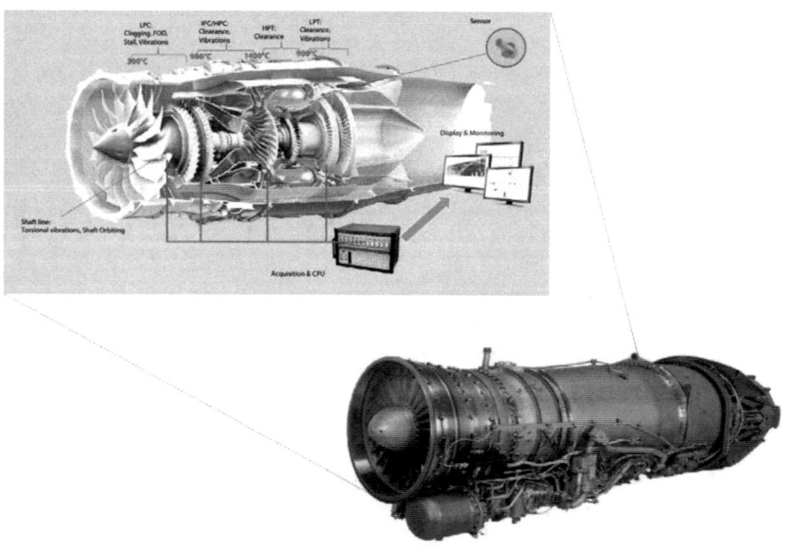

Figure 1. Installation points for vibration sensors.

The second group of tasks is about the assessment of the vibrational state of the engine and its elements in order to prevent malfunctions caused by vibrations (Figure 2). The vibrational state of the engine is defined the set of the parameters characterizing the vibration of the engine. The vibrational state can be determined for a given point the engine marked, a certain zone of diagram or the engine as a whole, as well as for a family of engines. The vibration state is determined for certain operating modes or for a set of engine operating modes, as well as for various operating conditions. Here are below the fault categories, which identify experts in the spectral characteristics of vibration signals.

Turbine rotor malfunctions: rotor imbalance, attachment failure, unacceptable runout of the rotor, unacceptable axial shift.

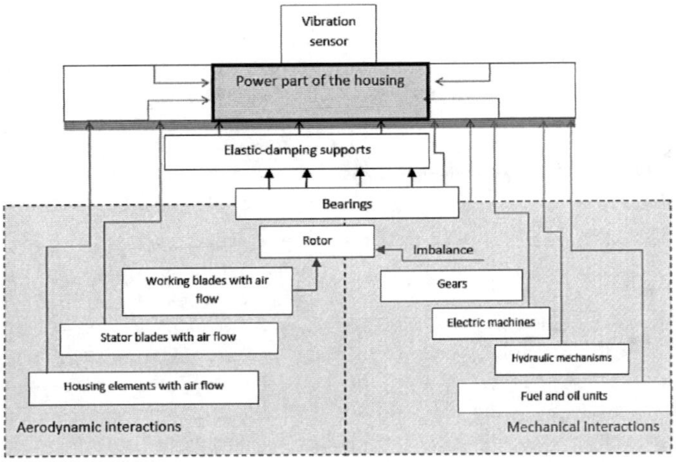

Figure 2. GTE vibration formation pattern.

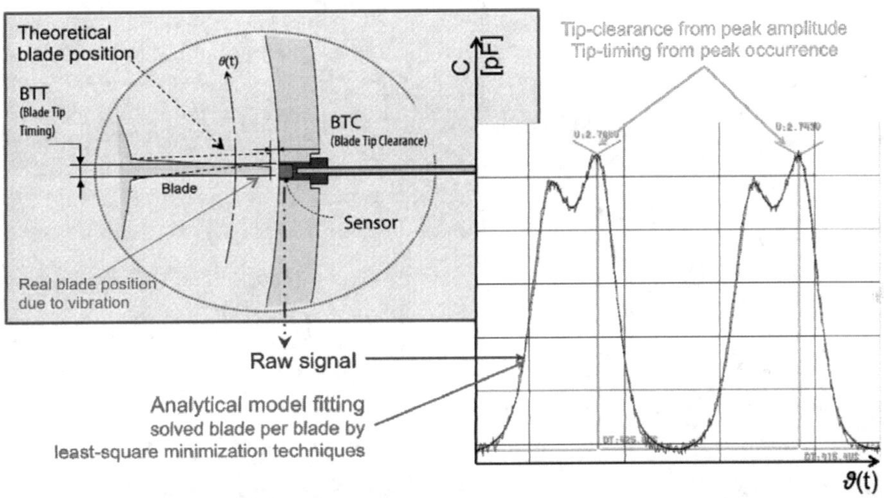

Figure 3. Measuring scheme of vibration accelerations using vibration sensors.

To assess the vibration velocity, the RMS of the vibration velocity is most often used — the root-mean-square velocity of the measured vibration velocity. The standards define a method of measuring the RMS — in the frequency range from 10 to 1000 Hz and a number of values of the RMS of vibration velocity: ... 4.5, 7.1, 11.2, ... — they differ by about 1.6 times. For units of different types and capacities, the value of norms of these series is set. When assessing the vibration state, one of the following parameters is:

RMS of vibration velocity in the frequency band 10–1000 Hz, Ve [mm/s];

RMS of vibration velocity in the octave frequency band, including the rotor speed, [mm/s];

Range of vibration displacement, [microns].

Evaluation of the vibration state of any node of the mechanism is carried out by comparing the maximum of the parameters measured of this node with the normalized limit value.

Due to this way of assessing the technical condition for most types of engines the following rules are true:

If Ve not greater than approximately 5 mm/s, the state of the engine is considered to be normal. The engine can be operated without restrictions;

If Ve exceeding 5 mm/s, but not exceeding circa 12 mm/s, the state of the engine is considered to be acceptable. The engine is operational, but needs repair. Operation of the engine is not recommended;

If Ve exceeding 12 mm/s, the state of the engine is considered to be dangerous. Do not start the engine and eliminate defects as soon as possible.

The maximum value of the range of vibration displacement can be estimated by the formula:

$$S_з \approx \frac{500 V_e\,[\text{мм/с}]}{f\,[\text{Гц}]} \qquad (1)$$

where f — turbine rotor speed.

Below, here is a spectrogram by time. Data was obtained by the SlamStick diagnostic software on the outside of the aircraft when it rose from 23,000 feet to 40,000 feet. The temperature dropped from 14°C to 31°C (58°F to -24°F) during the test.

Evaluation of equipment condition by the general level of vibration is one of the simplest forms of technical condition assessment. The maximum value is used as a diagnostic parameter (Figure 4), the list of conditions (normal, permissible, dangerous) is used as the technical state classes, and the algorithm for comparing the diagnostic parameter with the normalized limit value is used as diagnostic algorithms.

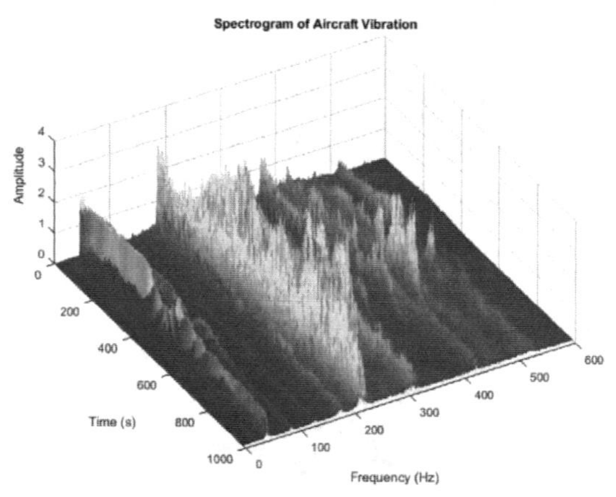

Figure 4. 3D spectrogram of an airplane engine obtained by SlamStick software.

The envelope spectrum diagnostic procedure usually consists of the following operations:

1) The resonance frequencies of the diagnosed nodes are determined (for example, the resonant frequency of the gear);

2) The characteristic frequencies of disturbances are determined (for example, the tooth frequency);

3) In the process of the mechanism determines the envelope spectrum. In this case, the band-pass filter (usually 1/3 octave) is tuned to the resonance frequency of the diagnosed node (gear);

4) In the spectrum of the envelope, they monitor the frequency components characteristic of this type of defect (tooth frequency).

An increase in the modulation coefficient of the characteristic harmonic indicates the development of a defect.

3. Spectral response troubleshooting method

Algorithms have been developed for the spectral processing of multicomponent vibration signals based on a harmonic wavelet transform [5–7] which are consistent with the proposed model and allow estimating signal parameters in the time domain. To simulate GTE malfunctions a rotary stand with vibration sensors was developed (Figure 5).

The basic functions used in calculating the wavelet coefficients of the harmonic wavelet transform correspond in shape to the individual sections (both local and extended) of the vibration signals under study, which allows the use of harmonic wavelets to approximate these signals. The developed algorithms take into account the features of the considered vibrational signals and are implemented in 2 stages. At the first stage, the signal is cleared of noise using a modified threshold processing of wavelet coefficients, the use of which does not lead to a shift in the desired process boundaries. At the second stage, by selecting sections with insignificant wavelet coefficients, the signal is cleaned from noise to estimate the time boundaries of the established vibrational processes.

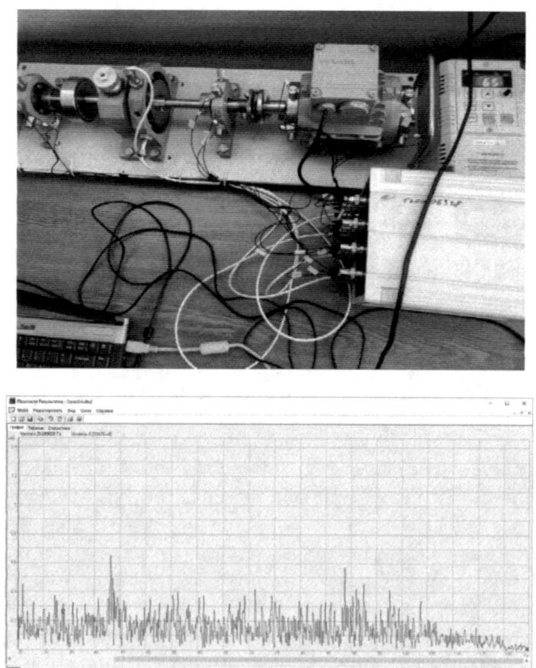

Figure 5. Simulator of GTE malfunctions and rotor spectrogram.

Due to the linearity property of the harmonic wavelet transform, the wavelet coefficients $a_{jvibr}(m)$ of the noisy signal $s(n)$ are represented in the form:

$$a_{jvibr}(m) = w_{jvibr}(m) + v_{jvibr}(m),\quad j = 0, ..., \log_2 N-1;\ m = 0, ..., 2^j-1, \tag{2}$$

where j is the level number of the wavelet decomposition; m is the number of wavelet coefficient; $w_{jvibr}(m)$ — wavelet coefficients of the useful signal; $v_{jvibr}(m)$ — wavelet coefficients of the noise presenting in the analyzed vibrational signal.

The peculiarity of this representation is that the wavelet coefficients $a_{jvibr}(m)$, corresponding to the analyzed vibration signal are complex. In this regard, their further threshold processing necessary to remove noise from the signal should be based on modified thresholds that take into account the distribution of noise over noise levels.

Due to the oscillating structure of the considered steady-state vibration processes (SSVP), the wavelet coefficients of the mixture of the steady-state vibration process and noise are approximately equal to the wavelet coefficients of the noise:

$$a_{jSSVP}(m) \cong v_{jSSVP}(m),\quad j = 0, ..., \log_2 N-1;\ m = 0, ..., 2^j-1, \tag{3}$$

where $a_{jSSVP}(m)$ are the wavelet coefficients of the mixture of the steady-state vibration process (SSVP) and noise; $v_{jSSVP}(m)$ — wavelet coefficients of the noise component of the steady-state vibration process.

Vibration values $w_{jvibr}(m)$ are defined as follows:

$$w_{jvibr}(m) = G\,|\,a_{jvibr}(m), \rho_j\,|, \tag{4}$$

where G is the nonlinear operator of the modified soft threshold processing of the wavelet coefficients $a_{jvibr}(m)$; ρ_j — the modified threshold values.

It was determined that the processing of noisy vibration signals set octave filters, harmonic respective wavelet transformation, the standard noise deviation (SND) for each successive level of decomposition is reduced in time compared to the previous one. The following expression was obtained for calculating the thresholds:

$$\rho_j = \frac{\sigma_q}{\sqrt{2}\cdot 2^{j/2}}\sqrt{2\ln N},\quad j = \log_2 N-2, \log_2 N-3, ... \tag{5}$$

where σ_q is the estimate of the standard deviation of the noise component at the thinnest level of decomposition (this level, due to the pronounced oscillating structure, contains the largest share of noise) with the number $q = \log_2 N-1$; the threshold ρ_j is applied to the real and imaginary parts of the wavelet coefficients.

The expressions for calculating the modified wavelet coefficients (taking into account the found thresholds) and the signal cleared of noise have the form:

$$\hat{w}_{jvibr}(m) = \begin{cases} a_{jvibr}(m) - \rho_j, & a_{jvibr}(m) > \rho_j; \\ 0, & \rho_j < a_{jvibr}(m) \le \rho_j; \\ a_{jvibr}(m) + \rho_j, & a_{jvibr}(m) \le \rho_j, \end{cases}\quad \hat{s}(n) = W^{-1}[\hat{w}_{jvibr}(m)], \tag{6}$$

where $\hat{w}_{jvibr}(m)$ — modified wavelet coefficients; $\hat{s}(n)$ — signal cleared of noise; W^{-1} is the inverse harmonic wavelet transform operator; $\rho_j = \frac{\sigma_q}{\sqrt{2}\cdot 2^{j/2}}\sqrt{2\ln N}$. The σ_q estimate is determined in accordance with the least squares method or median estimate (normalized median of the absolute deviations of the wavelet coefficients from their median).

After obtaining an estimate of the vibration signal, sections with insignificant wavelet coefficients at the thinnest level of the wavelet decomposition are selected. Modified wavelet coefficients $\hat{w}_{jvibr}(m)$ of the vibrated signal cleared of noise are considered insignificant if they satisfy the condition:

$$|\,\hat{w}_{jvibr}(m)| < \eta, q = \log_2 N-1;\quad m = (N/4)...(N/2)-1, \tag{7}$$

where $\eta \approx 10^{-5}...10^{-3}$.

Otherwise, the wavelet coefficients are significant and correspond to a transient or shock process. According to the results of segmentation in wavelet field it is generated an A_1 wavelet coefficients set at the subtle level of decomposition:

$$A_1 : \{\hat{w}_{jvibr}(m)\},\quad m \in [n_0, n_1] \cup ... \cup [n_j, n_{j+1}] \cup ..., \tag{8}$$

where $n_0, n_1, ..., n_j, n_{j+1}, ...$ are the boundaries of the established vibrational processes in the space of wavelet coefficients.

The found boundaries of insignificant sections are recalculated into the time domain, and thereby the vibrational signal is segmented.

To evaluate the signal parameters characterizing the potentially dangerous resonance phenomena and the energy properties of mechanical vibrations, a computational algorithm based on wavelet-smoothing [5, 6] Fourier periodograms [6] of steady-state vibration processes has been developed.

When developing the algorithm, it was taken into account that the processes under study have a continuous spectrum, an irregular structure, and distributed local features in the frequency domain, including the most important of them — resonance peaks. In addition, the algorithm allows us to use not only the samples of the signal in the time domain, but also the samples of its DFT (discrete Fourier transform) or the samples of its Fourier periodograms coming from the object

IOP Conf. Series: Materials Science and Engineering **714** (2020) 012014 doi:10.1088/1757-899X/714/1/012014

via telemetric communication channels as initial information for estimating the parameters of a multicomponent vibration signal.

4. Implementation using a neural network

Convolutional Neural Network (CNN) [7] is a class of models that successfully solve the problem of modified wavelet transform. Convolutional neural networks are part of the "deep learning" machine learning paradigm, which has proven itself very well in various tasks. The principle of its use is the sequential use of convolution operations and the choice of the maximum value (max-polling [8]).

At first, the neural network classifier was trained with tuned wavelet coefficients [9, 10]. Recurrent layers highlight the spectrogram feature space, which allows finding a suitable critical cause of the defect and displayed on the spectrogram.

At the output there will be a vector $K\{k_1, ..., k_n\}$; the coefficients kn are the correlation coefficients with clusters of spectrogram images of individual GTE nodes, produced the neural network (Figure 6).

Figure 6. Configuring neural network classifier layers and spectrogram input.

To train the layers of a neural network without a teacher, gradient minimization of error is used. Recurrent auto encoders allow you to work not with a single input vector, but with the time sequence of vectors. The number of vectors in the sequence is set by the parameter val_acc (validation data vector) and val_loss (loss data vector) (Figure 7) [11, 12].

After each stage of training, the quality of training achieved at this stage was checked. For this, sets of input data that were not involved in the training were presented at the network inputs [13, 14]. After that, the probability (share) of correct and false recognition (classification) of the input data sets was calculated and a table of the maximum probability of incorrect classification depending on the number of the training stage was compiled [15]. Training is considered successful if the reliability of the classification of faults is at least 0.95.

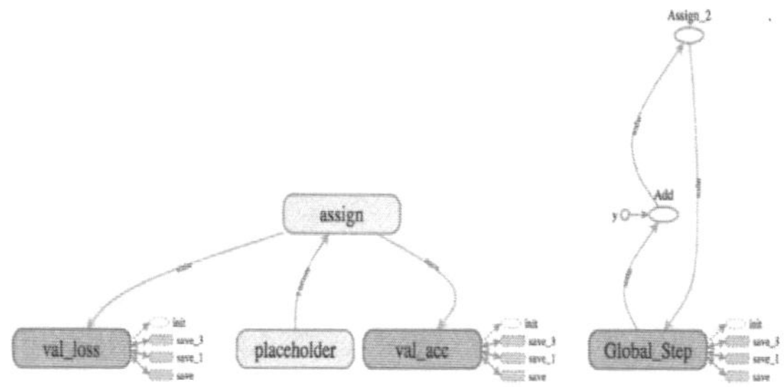

Figure 7. The scheme of training layers of the neural network classifier.

When considering Figure 4, four clusters with deviations from the norm can be distinguished. In two clusters, it is clearly seen that the dispersion will not exceed 0.2. Such clusters reveal groups of vibration accelerations in the spectrogram, which clearly indicate the effect of thermal expansion of the rotor.

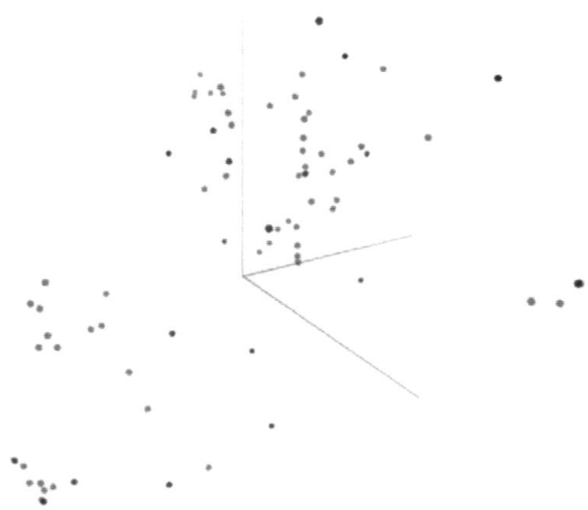

Figure 8. Visualization of vibration acceleration vectors in hidden space using t-NSE.

In Fig. 5, the vibration acceleration vectors are represented in a hidden space using t-NSE, it is also possible to distinguish 4 main clusters corresponding to critical vibration states. Two clusters closer to the center contain data sets corresponding to vibrations associated with thermal expansion of the rotor.

The method was tested in the analysis of the spectral characteristics of vibration parameters measured during the operation of the developed stand, simulating the malfunction of the gas turbine engine. For various turbine generator failures, vibration signals were obtained, to which the fast Fourier transform was applied. As a result, signal spectra (spectrograms) were obtained for each type of fault. Informative diagnostic features represent the amplitude of the spectrogram at given frequencies.

After each stage of training, the quality of training achieved at this stage of training was checked by presenting input data sets that did not participate in the training to the inputs of the simulation stand. After that, the probability (share) of the correct and false recognition (classification) of the input data sets was calculated and a table of the maximum probability of incorrect classification depending on the number of the training stage was compiled. Training is considered successful if the reliability of the classification of faults is at least 0.95.

To evaluate the accuracy of the forecast, we calculate the coefficient of determination of R^2 and the average absolute error (MAE). Mean Modulus of Deviation (MAE — Mean Absolute Error or MAD — Mean Absolute Deviation):

$$MAE = \frac{1}{m}\sum_{i=1}^{m} | a_i - y_i |, \tag{9}$$

where a_i is the predicted result, y_i is the actual result, m is the sample size, R^2 is the variance fraction of the dependent variable.

It was calculated the value of the reliability of detection of the predicted malfunction and the indicators of the accuracy of the forecast (table 1).

Table 1. The results of neural network classification and the fixed distribution of forecast errors.

№№	Fault signal	Validity of definition	Operand function	R^2	MAE
11	«Rotor imbalance»	0.95	HEAT DEFLECTION ROTOR GTE	0.86	730
22	«Rotor imbalance»	0.98	GTE ROTOR IMBALANCE	0.81	620
33	«Loosening the mounting of the support nodes»	0.98	TOUCHING IN THE OPPOSITE PART OF THE SUPPORT NODES	0.94	650
44	«Rotor imbalance»	0.96	ENGINE SHAFT ROTOR CURVING	0.86	620
55	«Damage to babbit loose leaves»	0.98	BEARING SHAFT IN BEARING	0.88	690
66	«Increased clearance of rotor bearing»	0.97	VIBRATION LEAP	0.71	870
78	«Increased clearance of rotor bearing»	0.99	BEARING VIBRATION MEASUREMENT CHANNEL MALFUNCTION	0.9	640

Based on the analysis of the reaction of the vibration analysis system, it can be concluded that the application of the module for forecasting the category of malfunction of a gas turbine engine is effective.

For this input, the reliability of the fault classification was calculated.

The reliability of the classification of malfunctions according to the parameters of the absolute vibration of the bearings of the GTE bearings was at least 0.95.

5. Conclusion

Thus, an approach to the analysis of aircraft GTE malfunctions based on spectral characteristics is proposed. For analysis, we used a modified wavelet transform apparatus using a neural network of the recurrent autoencoder type. As a result, the neural network has learned to identify clusters with critical vibrational states at a time interval of more than 24 hours. When new data appears, the prediction module will establish a data cluster with a dispersion of no more than 0.2 and determine the type of predicted malfunction.

This approach and methods have their limitations, studies of which can be the subject of further search and experiments, including those used with other non-destructive testing means of rotor assemblies [16].

Acknowledgment

This research was supported by the Ministry of Education and Science of the Russian Federation as part of the federal target program. Application number 2019-05-576-0001-290 "Development of a modern budgetary stationary system for vibration protection and vibration diagnostics of critical rotary units of distributed energy".

References

[1] Stone G C, Boulter E A, Culbert I et al. 2014 *Electrical Insulation for Rotating Machines* 2nd ed. (Hoboken NJ: Wiley)
[2] Ponce H O, Gott B and Stone G 2011 *Generator Stator Endwinding Vibration Guide: Tutorial* EPRI. Project Evaluation No 6382
[3] Shabaev V M, Kazantsev A S, Leontiev M K, Garanin I V and Karasev V A 2007 *Control. Diagnosis* **11** 18–27

[4] Kuchinskii K A 2010 *Pratsi Instytutu Elektrodynamiky Natsionalnoi Akademii Nauk Ukrainy* **27** 46–51
[5] Titko A I, Kuchinskii K A, Akhremenko V L and Osadchii E P 2006 *Pratsi Instytutu Elektrodynamiky Natsionalnoi Akademii Nauk Ukrainy* **3** 33–40
[6] Okhtilev M Yu, Sokolov B V and Yusupov R M 2006 *Intellectual technologies for monitoring and managing the structural dynamics of complex technical objects* (Moscow: Nauka)
[7] Kuznetsov D V and Shandybin M I 2007 *Power stations* 10
[8] Lee J Y and Dernoncourt F 2016 Sequential Short-Text Classification with Recurrent and Convolutional Neural Networks *Proceedings of NAACL-HLT* pp 515–20
[9] Nazolin A L and Polyakov V I *OOO NTC "Resurs". Vibroacoustic diagnostics and stator monitoring*
[10] Morozova T Yu, Bekarevich A A and Budadin O N 2014 *Control. Diagnostics* **8** 42–8
[11] Akimov D A, Rabotkin V D, Teryokhin I V and Redkin O K 2013 *Industrial control systems and controllers* **1** 33–8
[12] Ostroukh A, Nikonov V, Ivanova I, Sumkin K and Akimov D 2014 *Journal of Artificial Intelligence* **1** 1–12
[13] Akimov D, Krug P, Ostroukh A, Ivchenko V, Morozova T and Sadykov I 2016 *International Journal of Applied Engineering Research* **9** 6435–40
[14] Akimov D A 2011 *Industrial control systems and controllers* **5** 1–6
[15] Kulagin V, Ivanov A I, Kuznetsov Y M and Chulkova G M 2017 *Journal of Physics: Conference Series* **803** 012083
[16] Filinov V V, Mikaeva S A, Rodyukov M S and Filinova A V 2017 Modern architecture board information and control systems of heavy vehicles *Russian Technological Journal* **5(3)** 114–123 https://doi.org/10.32362/2500-316X-2017-5-3-114-123

Unified Approach to Solving the Trajectory Problem by Usage of Neural Maps

V S Kuvshinnikov[1] and E E Kovshov[1]

[1] JSC «NIKIMT-Atomstroy», Moscow, Russian Federation

E-mail: kuvshinnikov@ndtexpert.ru

Received xxxxxx
Accepted for publication xxxxxx
Published xxxxxx

Abstract

In modern robotics, there are many approaches to planning the trajectory of the portal manipulator. Methods based on the use of graphs, methods of potential fields, with the use of cellular decomposition, intelligent and others are used. Many varieties are based on similar mathematical apparatus. An important aspect of the trajectory search method applicability is convenience of chosen workspace and obstacles representation method. The approach proposed for manipulator path modelling implies a multi-level process, which includes periodic calculation and updating of invalid configurations set, considering the held cargo dimensions, a neural map creation and activation by the wave front propagation method etc. During experimental simulations with different activation functions, numbers of artificial neurons and synaptic connection implementation types, few methods of neural map activation have been considered, including flat and/or three-dimensional raster methods and wave propagation with static obstacles.

Keywords: neural map, trajectory planning, programming patterns, abstract model

1. Introduction

Currently, there are many approaches to the industrial manipulators path generation. They differ in related computations complexity, corresponding tasks specifics, result application flexibility, etc. Among others, there is a task to automate devices that do not require precise movement control and or implement high movement dynamics. Complete replacement with fully automated model is complicated by the fact that some of these devices, such as gantry cranes and overhead manipulators, are widespread, have a high cost and may require the entire manufacturing line suspension during assembly and installation process. Such manipulators have simple kinematics, work in principally consistent conditions, and solve a limited set of tasks, which allows to consider the simplified scheme automation for this class of devices.

2. Model explanation

This article deals with Cartesian coordinate system manipulators with 3 degrees of freedom (no kinematic redundancy) like some industrial overhead manipulators (OM) for example. Cartesian coordinate system is simple and combined with no kinematic redundancy allows making more illustrative and easier to understand visual map models.

2.1 Workspace representation

Manipulator kinematic scheme includes three controlled translational joints, used to make operational device (OD) follow the trajectory. Configuration space of the manipulator is discretized with resolution N, so that the coordinate in each of the joints j_1, j_2 and j_3 can take discrete values from range $\mathbf{q}_{x_j}$ $(x_j \in \{1, ..., N\})$, where $\mathbf{q}_{1_j}$ and $\mathbf{q}_{N_j}$ are the lower and

IOP Conf. Series: Materials Science and Engineering **714** (2020) 012015 doi:10.1088/1757-899X/714/1/012015

the upper constructive linear motion constraints of the joints. Manipulator configuration space model can be now represented as a set of N^3 vectors

$$DC = \{\mathbf{q}_a \mid a = 1...N^3\}, \tag{1}$$

where $\mathbf{q}_a = [q_{x_1} \ q_{x_2} \ q_{x_3}]^T$ — discrete manipulator configuration $x_j \in \{1, ..., N\}$, where $j = 1, ..., 3$; a — a one-dimensional index, taking a value of $a = N^2 \cdot (x_1 - 1) + N \cdot (x_2 - 1) + (x_3 - 1)$.

Considering that information about the set of obstacles $B = \{B_1, B_2, ..., B_m\}$ is sufficient, a set of collision free configurations can be determined $DC_f = \{\mathbf{q}_a \in DC \mid M(\mathbf{q}_a) \cap B = \varnothing\}$, enoting manipulator in configuration $\mathbf{q}_a$ as $M(\mathbf{q}_a)$ [1]. A set of safe configurations DC_s is defined as a known subset of $DC_f \supseteq DC_s$. A set of dangerous configurations consisting of the configurations that do not belong to DC_s, including B and $DC_f \cap DC_s$, is

$$DC_d = B \cup (DC_f \cap DC_s). \tag{2}$$

Dangerous configurations imply other manipulators reach of arm, maintenance staff pathways etc. Key trajectory selection strategy implies avoiding potentially dangerous manipulator configurations and only bringing the manipulator into safe configurations.

2.2 Simplified representation

To simplify workspace representation, to reduce the amount of stored data and the access time, certainly safe areas are specified by coordinate ranges. It's called a hierarchical grid representation. A potentially inaccessible area is represented as a set of corresponding OM configurations. At the same time, adjacent configurations are initially combined into three-dimensional groups, decreasing the amount of filtering area cells calculations. Initially all the configurations in potentially inaccessible area are considered dangerous. The size of the configurations group is selected as a multiple of the linear size of the base group. Sample size for a cubic configuration group:

$$N_g = n_g^3;\ n_g = l^i, \tag{3}$$

where N_g — cubic group's full size, n_g — cubic group's linear size, l — base group's linear size, i — group's scale. At $i = 0$, a single configuration of the manipulator is considered. Configurations group is a three-dimensional adjacent configurations convolution, which stores information about the corresponding workspace cell dangerousness. The convolution dangerousness is determined by the coefficient $d \in [0, ..., 1]$. Groups with $d = 0$ are completely safe and with $d = 1$ are completely dangerous.

2.3 Neural mapping

A special representation of manipulator configurations set — a neural map — is used for trajectory formation. Each group of configurations stands for a cell of workspace and corresponds to one neuron. The choice of the neural network type is determined by the problem: filtering the workspace cells by the cost of transition from each cell to adjacent cells, considering the weight function.

The scheme of the elementary network neuron is shown in figure 1.

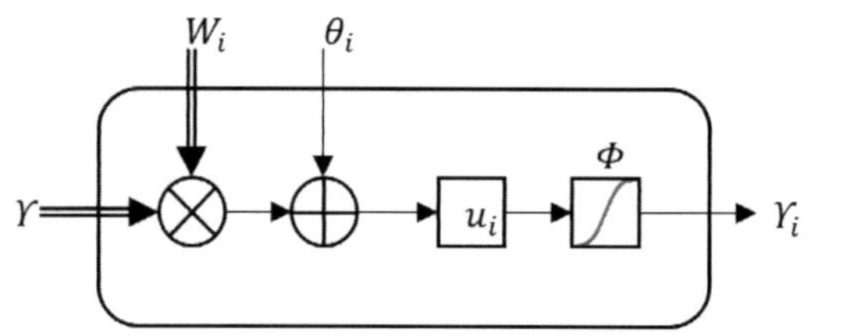

Figure 1. Elementary nonlinear processing neuron unit.

Neuron i is described by input vector $\mathbf{Y}$, weight vector $\mathbf{W}_i$, displacement vector θ_i, that form a sum network signal

$$u_i = \mathbf{W}_i \times \mathbf{Y} + \theta_i = \sum_{j=1}^{n} w_{ij} v_j + \theta_i. \tag{4}$$

The sum signal is processed by a nonlinear activation function Φ and formed the output signal of the neuron Y_i. Hyperbolic tangent, sigmoidal function and others can be used as activation function.

Neural mapping requires determining the following parameters: the order of the considered configuration groups; the size of the considered area; the type of weight function, and the topology of the adjacent cells relationships. Considered configuration groups scale determines the linear dimensions of the workspace cell, taken as a single position of the manipulator, which corresponds to one neuron. This parameter can be called the neural map resolution. The lower the scale, the more neurons will include a neural map, which will increase the trajectory sampling and allow better dangerous zone enveloping. Decreasing the scale increases the number of neural map neurons by a multiple of l^3, where l is the base configuration group linear size.

Piecewise given function is chosen as a weight function of the i-th and j-th neuron connection

$$w_{ij} = \begin{cases} 0, & \rho(q_i, q_j) \cdot k_f(x, y, z) = 0 \\ f[\rho(q_i, q_j)] & 0 < \rho(q_i, q_j) \cdot k_f(x, y, z) < r, \\ 0, & r < \rho(q_i, q_j) \cdot k_f(x, y, z) \end{cases} \tag{5}$$

where $\rho(\mathbf{q}_i, \mathbf{q}_j)$ — Euclidean distance between configurations, corresponding to i and j neurons;

$k_f(z,\phi)$ — axial modificators function that affects the transition costs and may be coordinate dependant as well; x, y, z — chosen configuration's coordinates; $f[\rho(\mathbf{q}_i,\mathbf{q}_j)]$ — decreasing function of the form $f(x)=1/x$; r — the radius of the neutron region i.e. how far each neuron affects other neurons through weight functions [2].

The process of weight distribution across the map is called map activation. The key idea is to gradually reduce weight as the activation wave propagates through the map, flowing around obstacles. So that the initial neuron with highest weight can be reached by path-of-steepest-ascent method.

Depending on r value, each neuron may have different number of synaptic connections. A 26-synapses scheme were used [3]. New neural map scheme were suggested to be a 6-synapses one.

3. 6-synapses scheme feature

New scheme has about 4 times less synaptic connections and takes less calculations for activation process. But due to its lack of connections, the scheme decides all 2D and 3D diagonal connections as 2-times and 3-times further than non-diagonal correspondingly. Thus the path mainly tends to be parallel to one of the axes and use manipulator drives one by one consecutively (figure 2a), which is not efficient. Neuron's weight equiscalar surface takes the form of octahedron. Figure 2b shows an equiscalar surface visualized with Cubix — interactive visualization interface [4].

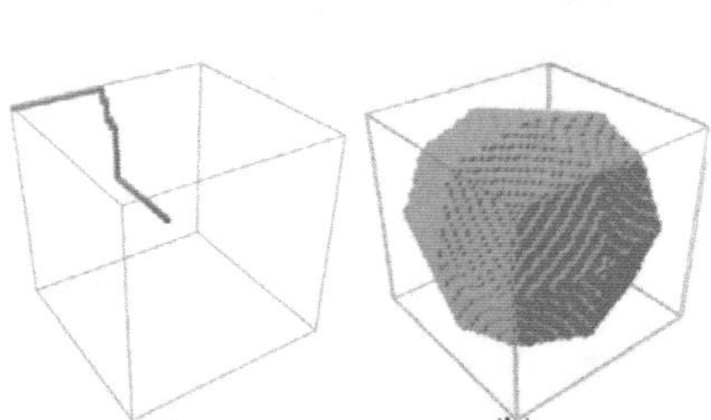

Figure 2. 6-synapses model results.

New activation function is needed to make trajectory more suitable for concurrent operation of drives, resulting in more spherical equiscalar neuron-weight surface.

One way is to consider diagonal connections as a single step $\sqrt{2}$ and $\sqrt{3}$ times further than usual. The activation function takes the following form:

$$\Phi = (1+\varepsilon\cdot\sqrt{j-1})/(1+\varepsilon\cdot\sqrt{n}), \quad (6)$$

where ε — minimal Euclidean distance between configurations, n — number of input synapses, j — input synapses deviation coefficient.

The other way is to consider diagonal connections as $\sqrt{2}$ and $\sqrt{3}$ steps of normal length. In this case the activation function takes the following form:

$$\Phi = [1/(1+\varepsilon)]^{\sqrt{j}-\sqrt{n-1}}. \quad (7)$$

Both functions turned out to overpower neuron activation level in some intermediate directions (figure 3a). It happens because of the grid model restrictions and neurons activation gradual accumulation. This can be partially corrected with an additional coefficient (figure 3b).

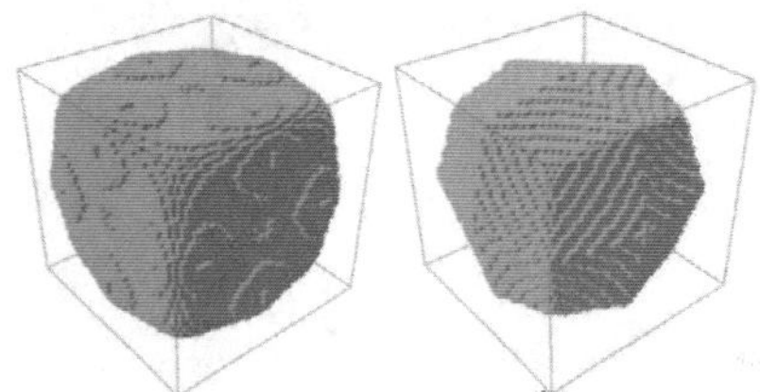

Figure 3. Equiscalar surface for new activation function.

Or another activation function can be chosen based not on geometrical calculations, but on required properties instead:

$$\Phi = \{1+\varepsilon\cdot[(n-1)/6]^j\}/(1+\varepsilon). \quad (8)$$

All three activation functions provide quite similar results, but the third one still has better computation efficiency and sphericity.

4. Anisotropic map issue

Neural map's specific feature is direction-specific activation modificators implementation capability. This however means that there are map activation sequence issue. If neurons are activated consecutively wave by wave — there will be an area where activated neurons with significantly different weights are met. In this case there should be an algorithm for neurons with higher weights to reactivate their lesser neighbours. Such algorithm turned out to be quite complicated and computationally expensive solution.

The better way is to make an activation queue. Currently activated neurons add their activation pending neighbours to a queue. Only the top weight neurons are processed during each turn of activation.

5. Implementation suggestions

Described method provides several advantages, such as that activated map needs little calculations to navigate, once activated, neural map only determines first or last trajectory point and allows to either reach it from any accessible point or vice versa. So neural map can be used to operate manipulators in case of large-series production without re-activation expenses. A complete set of pre-activated maps requires obstacles map, most essential locations coordinates, operated cargo weights and dimensions etc.

6. Conclusions

Although three dimensions neural map is resource intensive method because of consuming activation procedure, it still can be suitable for some tasks. Significant obstacles changes is a typical reason for map re-activation: the lesser the changes of obstacles and, at the same time, the more trajectory variability is needed — the reasonable the choice is. This approach provides a reliable way to navigate through even heavily cluttered workspace, still keeping widest variability opportunities, using a given set of once activated maps with low further navigation computational complexity.

Acknowledgements

The authors express their deep gratitude to the leadership of the research Institute of design and construction technology (NIKIMT) for their support throughout the entire R&D cycle.

We also thank our colleagues from the Department of information technology for their assistance during the modern software and hardware infrastructure creation and all essential spectrum for software development and numerous computational experiments in the conducted scientific research course.

References

[1] Lobareva L and Kozhevnikov M 2017 *Bulletin of Polotsk state University* **11** 43
[2] Lagoudakis M and Maida A 1999 Neural Maps for Mobile Robot Navigation *Proc. Int. Conf. on Neural Networks (IJCNN)* (Washington) pp 2011–6
[3] Kuvshinnikov V and Kovshov E 2018 *ITM Web Conf.* **18** 01004
[4] Bach B and Pietriga E 2014 Visualizing dynamic networks with matrix cubes *Proc. Ann. Conf. on Human Factors in Computing Systems (CHI 2014)* pp 877–86

Reliability Analysis of the Air Transportation Network when Blocking Nodes and/or Connections Based on the Methods of Percolation Theory

Sergey Lesko[1], Anton Aleshkin[1] and Dmitry Zhukov[1]

[1] Department of management and systems modelling, MIREA - Russian Technological University, Moscow, Russia

E-mail: Antony@testor.ru

Received xxxxxx
Accepted for publication xxxxxx
Published xxxxxx

Abstract

The paper shows that to study the reliability and fault tolerance of air transportation networks, methods of percolation theory can be used, in which any aviation transport structure can be represented as a random non-planar, incompletely connected graph (nodes are airports, arcs are airlines). In the theory of percolation, one can consider the solution of the problems of finding the shares of blocked nodes and blocked connections for networks with various random and regular structures, in which they decompose into unconnected areas. The share of blocked nodes (in the node problem) or connections (in the connection problem), at which the conductivity between two arbitrarily selected network nodes disappears, is called the percolation (flow) threshold. For the same structure, the values of percolation thresholds for the bond problem and the node problem have different meanings. The percolation threshold value depends on the average number of connections per network node (density), and is a criterion for its reliability, i.e. determines the percentage of blocked nodes and/or communications that the network will lose the necessary level of performance ability. The dependence of the blocking threshold (percolation) on the network connection density can be expressed mathematically. Using a map of a real aviation transport network, it is possible to determine the average number of connections per one node and then calculate the threshold value of its predetermined reliability value. If the reliability threshold needs to be increased, then the necessary number of additional links can be calculated.

Keywords: aviation transport network, network connection density, network percolation threshold, increasing reliability, increasing throughput

1. Introduction

Recently, to the study of flow balancing and ensuring the smooth operation of transport systems much attention has been paid. It should be noted that a significant number of works in this area are devoted to land transport [1–14] and there are few studies in the field of air transportation. First, the problem is that the number of vehicles is constantly growing, and the development of road infrastructure is late. As a result, the applied control models become obsolete and fail. Thus, the need arises either to search for new management tools, or to modernize the physical basis of the existing transport network.

For aviation, there is also the problem of increasing traffic with limited bandwidth available airports. All this also entails the need to develop models and methods for controlling and balancing air traffic flows, considering the fact that the infrastructure of air systems is significantly different from ground-based ones. In a generalized sense, it is necessary to solve the dynamic problem of the redistribution of flows (traffic) in transport networks, considering their topology. Moreover, if ground networks can be reconstructed due to relatively small investments (construction of flyovers and interchanges), then modernization of aviation infrastructure requires significant costs (expansion of existing and construction of new airports), compared to ground systems. Therefore, the main instrument for solving the problem of ensuring the smooth operation of airlines can only be changing command models.

Control and balancing of flows in transport networks is one of the most important problems of ensuring uninterrupted aviation traffic. Violations in the work lead to disruptions in the transportation of passengers and goods, which can be considered as a decrease in reliability, leading to significant financial losses.

It should be noted that local control, which is based on the statistically estimated characteristics of traffic flows, is not applicable in this case, because the result is provided by obtaining an estimate of the efficiency of the functioning of traffic flows at one transport hub, excluding neighboring ones. Guidance system is needed that will allow for optimizing the functioning of traffic flows in an area that includes many transport nodes. A change in control actions on one transport node inevitably causes a change in the characteristics of traffic flows on neighboring ones.

The bulk of all ongoing studies of transport traffic, its analysis and development of control models is aimed at solving emerging problems at the local level, without considering the entire transport network.

In our opinion, it is necessary to use a broader approach that considers the transport properties of the entire network. Ensuring the smooth operation of aviation transport networks is possible based on their analysis using the methods of percolation theory, which allows us to evaluate the conductivity of the entire network using the probabilities of blocking individual nodes or links.

2. Problem Statement

The transport network of modern aviation companies can have a very large, complex and branched structure (see, for example, Figure 1, which shows the United Airlines airline communications diagram taken from the resource: https://happynewyear2018cards.com/united-airlines-flight-map/united-airlines-flight-map-stunning-us-airways-flight-map-europe-refrence-united-airlines-route-map/, which can be represented as a random non-planar, incompletely connected graph (nodes — airports, and ribs — transport lines, see Figure 2).

Figure 1. The scheme of airline routes of the company United Airlines.

Not the planarity and incompleteness of connectivity of the air transportation networks is obvious, but the randomness of their graph is not in the randomness and randomness of the movement of airplanes between airports, but in the fact that an arbitrarily chosen node can have a different number of connections (central nodes (for example, Chicago, San Francisco or Denver there are significantly more than the rest).

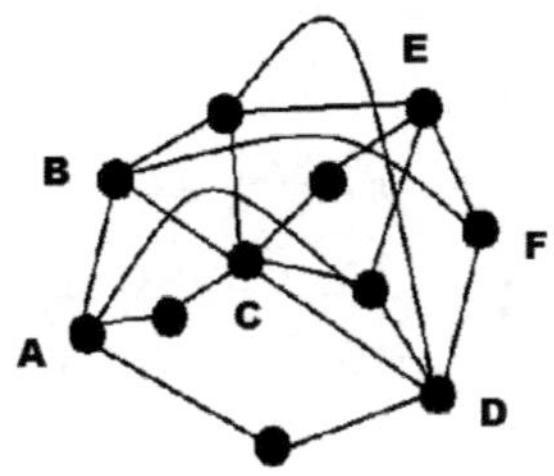

Figure 2. Representation of the air traffic network in the form of a random network structure.

When simulating air traffic, it is necessary to consider the dynamics of changes in air traffic congestion (daily change in the flow rate) and the fact that all elements of the transport network graph (nodes and edges) have different characteristics (throughput).

If we analyse the percolation model of the transport aviation network, we can ensure the necessary flow balancing and reliability of the entire structure, even though its individual elements can be blocked due to overload or denial of service. In this case, by ensuring operability and reliability, we mean that between any two arbitrary network nodes (not necessarily central ones) there is at least one free path from the unblocked elements of the graph.

In the theory of percolation (probability theory on graphs), the solution of the node problem and the connection problem [15–18] is studied for networks with different, both regular and random structures. When solving the problem of bonds, they determine the proportion of bonds that must be broken so that the network breaks up into at least two unconnected parts (or vice versa, the fraction of conductive bonds when conductivity arises). In the problem of nodes, the fraction of blocked nodes is determined, in which the network breaks up into unconnected clusters within which connections are preserved (or vice versa, the fraction of conductive nodes when conduction occurs). The fraction of non-blocked nodes (in the node problem) or unbroken bonds (in the connection problem), at which conductivity arises between two arbitrarily selected network nodes, is called the percolation (leakage) threshold. For the same structure, the values of percolation thresholds for the bond problem and the node problem have different meanings. Note that when a node is blocked, all its connections are blocked, while when a connection is blocked, only one connection between neighbouring nodes.

Using the concept of shares of blocked nodes or links is equivalent to the concept of the probability of a randomly selected node (or link) being in a blocked state. Therefore, it can be accepted that the percolation threshold value determines the probability of passing through the entire network as a whole if some part of its nodes (or links) is blocked (excluded), i.e. the average probability of blocking a single node (or link) is given.

Percolation loss can be defined as the proportion of blocked network nodes, in which the entire network loses its conductivity.

Reaching the percolation threshold in the network corresponds to a cluster in which there are connections between any of its arbitrary nodes. A so-called endless or contracting conducting cluster is formed.

For structures of finite size, conductivity can occur at different proportions of conductive nodes (or bonds, see Figure 3). However, if the network size (the number of nodes L) is directed to infinity, then the region of transition to the conducting state becomes compact (see Figure 3, curve I for a small-sized structure, II is an infinite network).

It should be noted that terrestrial transport structures can be represented as random, almost planar, incompletely connected graphs, while air ones do not have planarity. It is important that the percolation thresholds of planar and non-planar networks at the same connection density differ significantly [19–22].

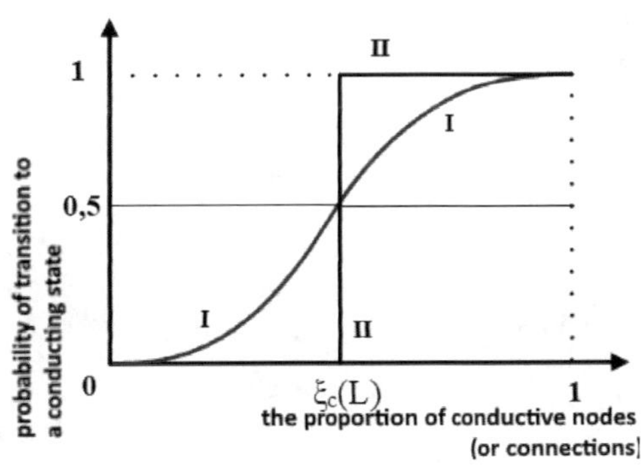

Figure 3. The probability of occurrence of percolation, depending on the magnitude of the fraction of conductive nodes (or bonds).

For structures of finite size, the percolation threshold $\xi_c(L)$ can be determined from a given value of the probability of the entire network transitioning to the conducting state. In Figure 3, this probability is chosen equal to 0.5.

However, the required value of the reliability value of the entire network as a whole may be set, for example, the value of 0.95 or 0.99 (then the percolation threshold will correspond to the specified criterion for the reliability of the network), i.e. it is possible to determine at what proportion of blocked nodes and/or communications the network as a whole will lose the necessary level of reliability.

It should be noted that in the analysis of reliability, central nodes or links can be assigned increasing weighting factors, thereby increasing their contribution to ensuring the overall reliability of the network.

3. Percolation properties of random non-planar, incompletely connected networks

The main problem in the study of the percolation properties of network structures having a random structure is that at present there are no general analytical methods for this, and their study is possible mainly only with the use of computer modelling methods.

3.1 Network percolation threshold calculation algorithm

To determine the percolation threshold of networks, the following algorithm can be used [19]:

1. First, build a random non-planar, incompletely connected network. Next, randomly select two network nodes A and B, considering the limitation that there must be at least one intermediate node between them.

2. Set the probability of blocking a single node (in the task of nodes) or communication (for the problem of communications) and randomly block the proportion of network nodes (or communications) equal to this probability.

3. Check the presence of at least one "free" path (path from not excluded nodes or links) from node A to node B. If there is no free path (the number of "free" paths is 0), write 0. Otherwise, write 1.

4. Increase the value of the probability of blocking a single node (in the task of nodes) or communication (for the problem of communication) by a certain value. Then randomly block the share of network nodes (or links) equal to a given probability value. Next, determine which specific network nodes were excluded.

5. Return to step No. 3 until all nodes of the network are enumerated.

6. Return to point No. 2 and perform points No. 3 – No. 5 — Q times (for example, several hundred times). From the first to the last steps (in cases where the entire network is blocked), for all experiments. Find the number of realizations for which at least one "free" path was found (we call this number n). For example, at $h=18$ step in 8, 12, 19, 56, 58, 76, 80 and 89 experiments from Q there was at least one "free" path, then the number $n(5)=18$ (8 is the total number of "free" ways). For each step, we find $\bar{p}(h)=n(h)/Q$, where h is the step number. Calculate the average cluster size of excluded nodes, the number of such clusters, etc. (for all N experiments at each step). The average cluster size can be defined as the ratio of the sum of all the average values obtained at this clusterization step (for all Q experiments) to the total number of Q experiments. For an explanation, consider an example. Suppose that at $h=6$ step in the 1st experiment, 4 clusters having a size of 15 nodes each were obtained, in the 2nd — 3 clusters, in the 3rd — 2, etc. in the 100th — 20. Then the average number of clusters having a size of 10 blocked nodes will be: $(4+3+2+5)/100$.

7. Next, return to the implementation of paragraph No. 1, and repeat the execution of steps No. 2 – 6 another W times. For each of the W experiments, we find the quantity $\bar{p}_w(h)=n(h)/Q$. The w index determines which of the W tests we are considering.

8. After the simulation is completed, for each of the h steps find the quantity $\langle\bar{p}(h)\rangle=\sum_{w=1}^{W=100}\bar{p}_w(h)/W$ — is the average value of the probability of passing through the network in general, for non-blocked nodes (or links for the task of blocking links) at each of the steps (taking into account various possible path configurations).

9. Find the magnitude of the percolation loss, which is equal to the fraction of blocked nodes (or connections) at which the network conductivity disappears (calculated by the formula: unit minus the share of conductive nodes, in the problem of nodes (or connections, in the connection problem)).

Performing calculations using this algorithm allows you to obtain an array of data for the dependence of the average value of the probability of passing through the network $\langle\bar{p}(h)\rangle$ as a whole on the share of blocked network nodes (or connections for the task of blocking connections), with a different average number of connections per node (density network).

Determination of the dependence of percolation thresholds of networks on their density (average number of links per node).

The results of numerical modelling and calculation of percolation thresholds for random non-planar, incompletely connected networks with different average number of links at each node for tasks of blocking nodes and links are presented in Table 1 [20–22]. In this case, to determine the percolation threshold, a reliability value of 0.5 (50%) was chosen. However, we note again that you can set any desired value of the reliability value of the entire air traffic network, for example, the value of 0.95 or 0.99 and determine the percolation threshold for a given value of reliability.

The data presented in Table 1 are linearized well in the coordinates: $\ln P(x)$ is the natural logarithm of the percolation threshold depending on $z=1/x$ (the reciprocal of the average number of bonds x (network density) per node), see Fig.4. The points in Figure 4 indicate the experimental data, and the solid lines correspond to linear equations.

In the task of blocking nodes, the dependence of the natural logarithm of the percolation threshold $\ln P(x)$ on the reciprocal of the network density $(1/x)$ can be described by the equation:

$$\ln P_{node,unreg}(x) = -\frac{1.70}{x} + 0.04 \qquad (1)$$

with the value of the correlation coefficient of numerical data and the equation of linear dependence equal to 0.97 (see line 1 in Figure 4). In the relationship problem:

$$\ln P_{bond,unreg}(x) = -\frac{6.74}{x} - 0.20 \qquad (2)$$

with the value of the correlation coefficient of numerical data and the equation of linear dependence equal to 0.99 (see line 2 in Figure 4).

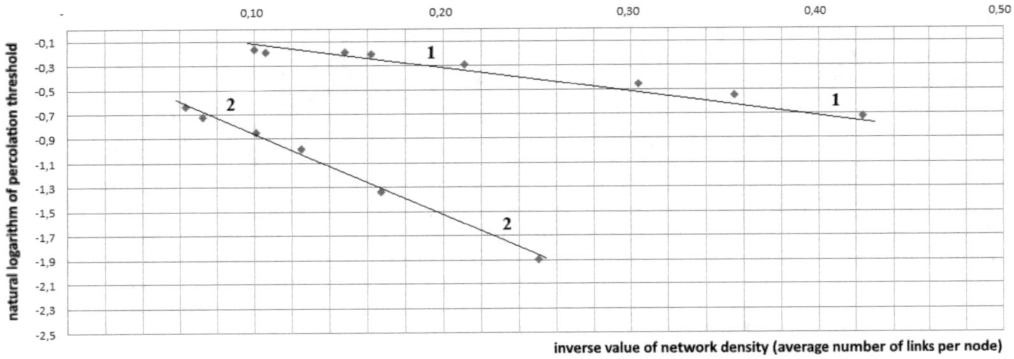

Figure 4. The dependence of the logarithm of the percolation threshold in a random non-planar, incompletely connected network on the reciprocal of the average number of connections (network density) per node (line 1 for the problem of blocking nodes and line 2 for the task of blocking connections).

Table 1. In parentheses shows the values of the reciprocal of the bond density and the values of the natural logarithm of the percolation loss threshold value.

Type of the criterion task	Average number of links per network node	The share of blocked nodes (or connections in the connection problem), in which the conductivity disappears in the network ($\xi_c(L)$ — percolation threshold)
The criterion task of blocking nodes	2.36 (0.424)	0.485 (–0.724)
	2.82 (0.355)	0.575 (–0.553)
	3.29 (0.304)	0.635 (–0.454)
	4.75 (0.211)	0.750 (–0.288)
	6.17 (0.162)	0.815 (–0.205)
	6.75 (0.148)	0.825 (–0.192)
	9.41 (0.106)	0.830 (–0.186)
	10.02 (0.100)	0.850 (–0.163)
Link Blocking Criterion Task	3.99 (0.250)	0.150 (–1.897)
	5.99 (0.167)	0.263 (–1.336)
	7.97 (0.125)	0.375 (–0.981)
	9.93 (0.101)	0.427 (–0.851)
	13.86 (0.072)	0.485 (–0.724)
	15.79 (0.063)	0.530 (–0.635)

3.2 Discussion of the results

Equations (1) and (2) make it possible to evaluate fault tolerance and compare random non-planar, incompletely connected networks with different connection densities, with a given value of reliability. In this case, these equations make it possible to give an estimate of operability for a value of the reliability level of 0.5 (50%), which is very small. However, can be set any desired value of the level of reliability of operability, for example, 0.95; 0.99 or 0.999. Then it is possible to build random non-planar, incompletely connected networks with different connection densities, and using the algorithm described above, calculate the values of their percolation thresholds at a selected level of reliability. Further, using the results obtained, it is possible to linearize the data and determine the corresponding linear dependences like equations (1) and (2).

Using equations (1) and (2), compare the reliability of air transportation of any randomly selected sufficiently large airlines (for example, the American company United Airlines and the Indian Jet Airways).

The United Airlines fleet consists of 787 aircraft (for 2019: http://www.airlines.id1945.com/1usr/united-ru.htm) and it is one of the four largest US airlines (in terms of fleet size it is in third place in the world, by the number of passengers carried in the fourth (at the end of 2017)). Jet Airways (data taken from the resource: https://india.ru/jetairways/jetairways_map.html) is the second largest airline in India (now its fleet has about 120 aircraft: https://www.skyscanner.ru/airline/airline-jet-airways-9w.html).

The United Airlines airline communications network, the diagram of which is shown in Figure 1, has a link density per node of about 7.4, and the graph of the Jet Airways airline network (see Figure 5) is about 3.2 (there are many hanging peaks).

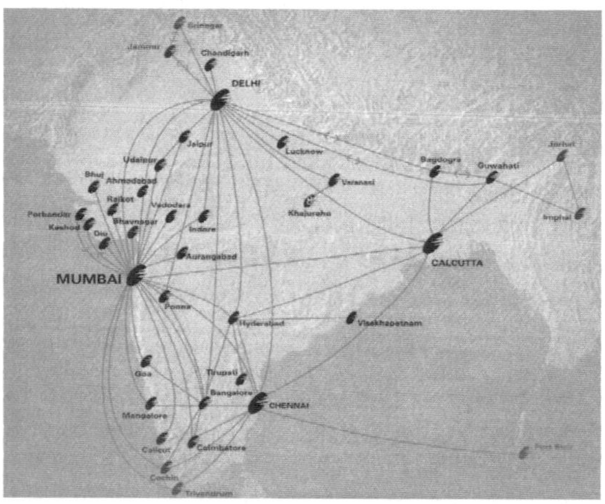

Figure 5. The scheme of aviation routes of the Jet Airways company.

Using equation (1), for the United Airlines aviation network in the nodes problem, the percolation threshold value (with a reliability level of 0.5) is 0.827, i.e. with a share of blocked nodes of 0.827 with a reliability level of 0.5 (50%), the airline's aviation network will lose its functionality. For the aviation network of Jet Airways, we get — 0.612. Thus, the reliability of ensuring the operability of United Airlines in the task of nodes is 1.35 times greater than that of Jet Airways ($0.827 / 0.162 = 1.35$).

Using equation (2), for the United Airlines aviation network in the communications problem, the percolation threshold value (with a reliability level of 0.5) is 0.329, i.e. with a share of blocked communications of 0.329 with a reliability level of 0.5 (50%), the airline's aviation network will lose its functionality. For the aviation network of Jet Airways, we get — 0.099. Thus, the reliability of United Airlines' operability in the communications task is also greater than that of Jet Airways by 3.32 times ($0.329/0.099 = 3.32$).

With an increase in the reliability level of performance capability to 99.99%, the share of blocked nodes or links at which the percolation threshold of this reliability level is significantly reduced. However, the general tendency is that the higher the network density, the higher the percolation threshold for loss of reliability is maintained.

4. Conclusion

To study the reliability and fault tolerance of aviation transport networks, one can use the methods of percolation theory. In which any transport structure can be represented as a random non-planar, incompletely connected graph.

In the theory of percolation, one can consider the solution of the problem of blocking nodes and the problem of blocking connections for networks with different structures. For the same structure, the values of percolation thresholds in solving the connection problem and the node problem have different meanings. The percolation threshold value depends on the average number of connections per network node (density), and is a criterion for its reliability, i.e. determines the percentage of blocked nodes and/or communications that the network as a whole will lose the necessary level of performance for a given value of reliability.

The dependence of the blocking threshold (percolation) on the network connection density can be expressed mathematically. This allows using the map of a real transport network to determine the average number of connections per one node and then calculate the value of its blocking threshold at a given level of reliability. If the blocking threshold needs to be increased, then the necessary number of additional links can be calculated.

5. Appreciation

The results presented in this paper were obtained within the programme of choosing scientific projects performed by scientific laboratories teams, according to the letter of the Ministry of Education and Science of Russia, 3 July 2019 No. MH-1037/AM. Project 2019-1393 "Developing new models for analysing and forecasting the dynamics of stochastic processes in sophisticated systems considering self-organisation and the presence of memory".

References

[1] Briani M and Cristiani E 2014 *Networks and Heterogeneous Media (NHM)* **3** 519–52

[2] Hui M, Bai L, Li Y and Wu Q 2015 *Mathematical Problems in Engineering* 20–7

[3] Ahn G -H, Ki Y -K and Kim E -J 2014 *IET Intelligent Transport Systems* **2** 89–95

[4] Poole A and Kotsialos A 2015 *Swarm intelligence algorithms for macroscopic traffic flow modelvalidation with automatic assignment of fundamental diagrams* (School of Engineering and Computing Sciences: Durham University)

[5] Guo J, Chen F and Xu C 2017 *Mathematical Problems in Engineering* 125–35

[6] Jiang R, Jin C and Zhang H 2017 *Transportation Research Procedia* **23** 157–73

[7] Danchuk V, Bakulich O and Svatko V 2017 *Procedia Engineering* **187** 425–34

[8] Pun L S C, Albert C and Chan W F 2016 *Travel Behaviour and Society* **3** 71–7

[9] Baranovskaya T P and Pavlov D A 2016 *Polythematic network electronic scientific journal of Kuban state agrarian University* **120** 1686–705

[10] Pavlenko P F 2014 *Problemy avtomatiki i upravleniya* **27** 92–7

[11] Trubitsyn V A and Golub D I 2013 *Bulletin of the North Caucasus Federal University* **2** 89–92

[12] Vlasov A A and Chushkina Z A 2014 *Regional architecture and construction* **4** 152–6

[13] Zyryanov V V 2013 *Energy and resource saving: industry and transport* **21** 71–4

[14] Filippova D M, Chernyago A B and Slobodchikova N A 2013 *Bulletin of Irkutsk state technical University* **9** 172–6

[15] Grimmet G 1999 *Percolation* (Berlin: Springer-Verlag)

[16] Sahimi M 1992 *Applications of Percolation Theory* (London: Tailor & Francis)

[17] Stauffer D and Aharony A 1992 *Introduction to Percolation Theory* (London: Tailor & Francis)

[18] Feder J 1988 *Fractals* (Plenum Press: New York, London)

[19] Zhukov D O, Andrianova E G and Lesko S A 2019 *Symmetry* **11** 920

[20] Zhukov D, Khvatova T, Lesko S and Zaltsman A 2018 *Technological Forecasting and Social Change* **129** 297–307

[21] Zhukov D O, Khvatova T Yu, Lesko S A and Zaltsman A D 2018 *Informatics and its applications* **2** 90–7

[22] Khvatova T Yu, Zaltsman A D and Zhukov D O 2017 Information processes in social networks: Percolation and stochastic dynamics. *CEUR Workshop Proceedings 2nd International Scientific Conference "Convergent Cognitive Information Technologies", Convergent 2017* **2064** pp 277–88

Printed Load Cells on Clothing Made of Fire-Protective Fabric

A V Lozitskaya[1], A P Kondratov[1], V A Baranov[1] and E P Cherkasov[1]

[1] Moscow Polytechnic University, Moscow, Russia

E-mail: a.p.kondratov@mospolytech.ru

Received xxxxxx
Accepted for publication xxxxxx
Published xxxxxx

Abstract

The paper represents the trial results of nonwoven fabric stencil-screen printed gauge indicators that can be made on other two types of fabric. They can be applied either for specific producing of pilot and test pilots' uniform or for airport emergency and fire officers clothing. The research being carried out, volt-ampere characteristics of conductive belt layer and electrical resistance scale effected by distortion were identified. It was also revealed that aramid fiber fabric sensors were to be applicable for creep stress measurement at temperatures of 20–200 C.

Keywords: printed electronics, polymer strain gauge sensor, elongation, printed sensors for flexible elements, fire-protective printed gauge, temperature gauge

1. Introduction

It should be noted that flex strain and temperature gauge indicators are the essential members of cutting-edge multifunctional uniform of test pilots, pilots, airport emergency service and fire officers. Besides this, flex strain and temperature gauges are widely used in aviation robotic equipment and luggage handling units. Therefore [1], artificial intelligence and robotics industrial sector have a high priority as the areas for these kinds of indicators. The supreme task of the devices and the clothing with special gauges is a real time monitoring of human movements, strains, stresses and the temperatures of the fabric [1–4]. The strain gauges are to be extremely flexible, tensile and adaptable to the fabric used [2]. One must emphasize that strain-tension gages seem to be one of the most significant ones among all available devices on the market due to their cost–effective and functional properties [1]. They are not widely used as sensor substrates of nonwoven fabrics [5–7]. It is essential that strain properties related to the structure and fiber mixture [5, 6, 8] and bonding were scrutinized without electrical properties of conductive belt. In addition, wearable electronics was not taken into account as well.

The experts [9] believe that the gages' performance (the variation of electrical resistance under strain condition) is specific on nonwoven fabrics, nonconvertible damage on the first cycle of strain including marginal and major forces can occur. The changes of pseudo-unbroken layers of carbon nanotubes are demonstrated on the example of applying filtering through nonwoven fabric consisted of nanotubes, cyclic stretching following as well. The net structure of the fabric keeps the operating conditions of the gages under relatively intensive and considerable elongations.

To sum up, this paper deals with a simple and economically-effective approach to make printed gauge indicators with the distortion versatility by means of screen printing for two types of fabric at special clothing production.

2. Subjects and methods

Nonwoven sheets and two types of fabrics are chosen as a gauge indicator substrate. They are:

- Polypropylene fiber sheets which are called spanbond (density 60g/sm)

- Needled fabric made of polypropylene fibers
- Cotton fiber non-combustible fabric for clothing inner layer branded Frol 280a (Fig.1). It contains non-combustible infiltration agent of quintavalent poly-condensate comprising phosphonium and nitrogen-containing compounds branded AFLAMMIT ®SAP.

Flame resistant fabric for clothing outer layer made of aramid fiber mixture branded FlamFort W280 (Fig.2).

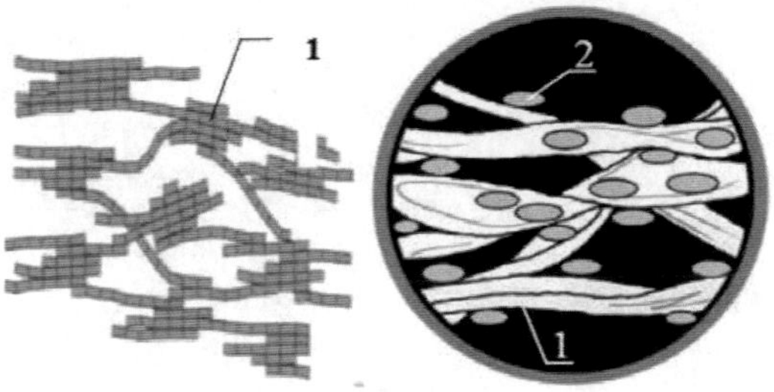

Figure 1. Cotton fiber structure implied for fire-proof clothing inner layer: 1 — cellulose fibers; 2 — antipyrene particles.

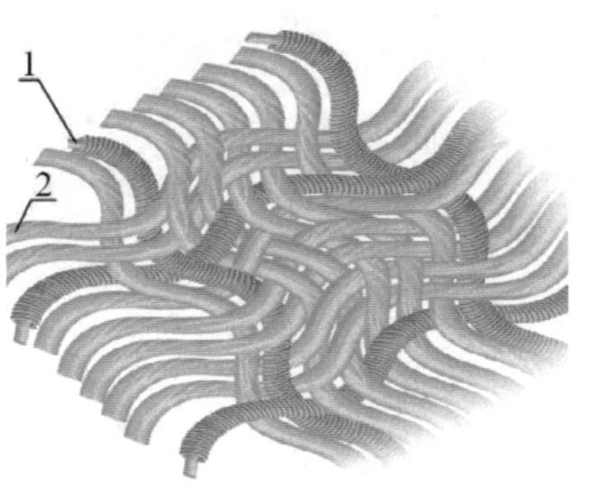

Figure 2. The wrath in scene of fire - proof polyaramid fabric implied for outer clothing layer: 1 — polymethaaramid fibers; 2 — plolyparaaramid fibers.

Fabric and nonwoven gauge indicators shaped as 10mm stripes were produced by printing plate stencil with net ruling 120 lpi. Electrically conducting stencil films Sun Chemical containing graphite dispersion and amorphous carbon were applied for gauge indicators' samples.

The samples were laboratory prepared at the temperature of 22±2®C being under conditioning and relative humidity of 45–50% [11].

The electrical resistance circuit with gauges' stress and strains in 50–100 Hz dc-and-alternating current network is attached to the article.

The stress-and-strains characteristics were identified in accordance with State All-Union Standard 15902.3-79. The gage resistance to increased air temperatures by means of sample exposure in temperature-cabinet for 20 minutes under RH 90–95% with consequent resistance following the methodology was noted [11].

3. Results and discussion

The initial and processed by printing gauge indicators on fabric samples were put to the mechanical tests on stress and strains. On examining the sheets, a strong effect in stress and distortion of electrically-conductive film was not detected.

The fabric distortion up to relative elongation 30–40% longwise and against the grain occurs. It takes place under mitigating stress caused by fibers weaving. (Fig. 2).

Electric and mechanical properties of flex gauge indicators are determined by homogeneity and layer printing composition on offsets. The surface texture is repeated. The samples of gauge indicators were manufactured on nonwoven polypropylene sheet, they were produced for comparison and determining the effect of surface texture on electric conductivity of the flex gauge indicators. The samples are varied greatly by microstructure and surface homogeneity.

On Figure 4 volt-ampere characteristics of printed resistors are shown, the alternating current frequency is 100 kHZ on the different density nonwoven fabrics substrates. It could be seen that electric resistance of the print on the thin and uneven thermal bonded sheet Spanbond exceeds twice the resistance of the same size needled nonwoven material. (2)

Volt-ampere characteristics of all samples with woven substrates are determined by linear characteristics, which makes them possible to be used as stable elements of electronic circuits formed by polygraph and combined methods. The printing on one type bonding sheets was implemented to obtain fire-p roof gauge resistors, the ones having similar structure and homogeneity.

Table 1. Distortion and strength of gauge indicator prints.

Direction of strength tests	Longwise		Across	
Parameters	No composition	With composition	No composition	With composition
Relative elongation, %	57	60	44	45
Strength limit, MPa	21,8	22	13,7	14

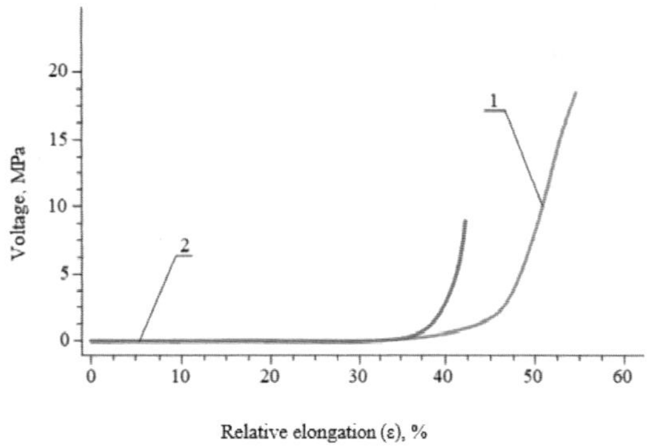

Figure 3. Technical diagram of aramid fabric strain, applied for outer clothing layer. 1 — longwise 2 — against the grain.

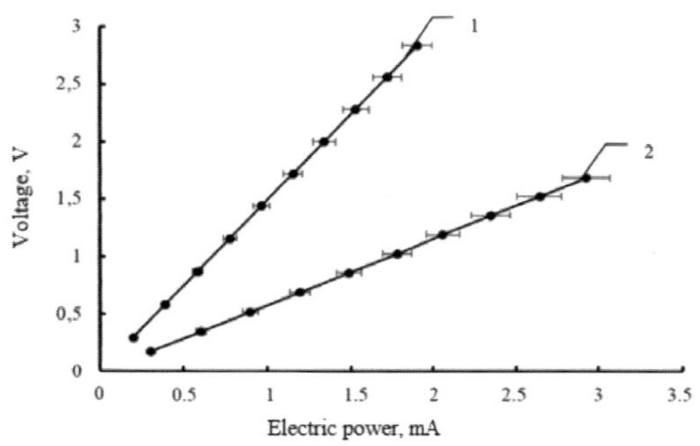

Figure 4. Relation of electric voltage to electric power on printed resistors, produced on non-woven sheets with the density of 1 — $60 g/cm^2$ and 2 to $60 g/cm^2$.

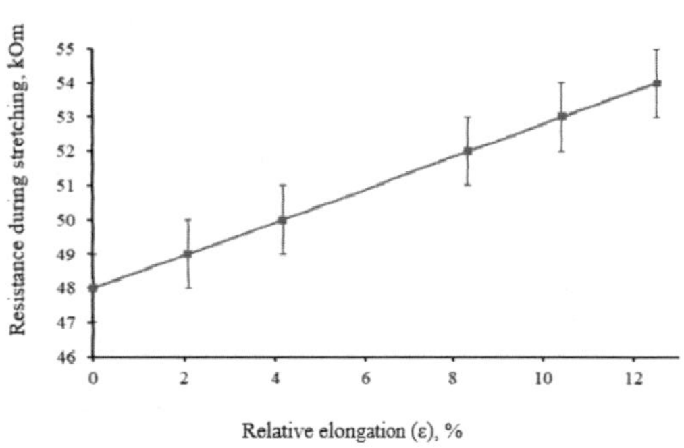

Figure 5. Relation of electrical resistance of printed on fire proof cotton fiber resistor to strain distortion (ε). $\Omega = 2E - 15\varepsilon^2 + 0.48\varepsilon + 48$ ($R^2 = 1$).

Table 2. Electrical resistance of gauge indicator prints to D.C (direct current)

The direction of the trial	Longwise				Against the grain			
Temperature, °C	20	50	100	200	20	50	100	200
Electrical resistance, Ω	550	600	550	600	450	500	450	500

4. Conclusion

Overall, the possibility to obtain the tension and distortion gages by means of screen printing on nonwoven fabric and other two types of fabrics are shown in this article. This technology can be used for producing the uniform for test pilots and pilots as well as for emergency service and fire officers. In addition, the stability of electrical characteristics in the range of 20–200 C was established. Besides that, the benefit of this type of printed sensors is the prospect to place them at any part of clothing as well as the variation of the voltage form and the voltage limits measured in composition with industrial production.

References

[1] Nag A, Menzies B and Mukhopadhyay S C 2018 *Sensors and Actuators A* **276** 226–36
[2] Cao J, Qin L, Liu J, Ren Q, Foo C, Wang H, Lee HP and Zhu J 2018 *Ext. Mechanics Lett.* **21** 9–16
[3] White E L, Case J C and Kramer R K 2017 *Sensors and Actuators A* **253** 188–97
[4] Nehir S O, Keskin A, KheaD and Onal C D 2015 *Sensors and Actuators A* **236** 349–56
[5] Advances in Technical Nonwovens ed. George Kellie 2016 (Woodhead Publishing)
[6] Omrani F et al. 2016 *Composites Part B* 116471–48
[7] Dal Pra I et al 2005 *Biomaterials* **26** 1987–99
[8] Savel'ev M A, Komarova L Y and Kondratov A P 2016 *Izv.Vys.Uchebn. Zaved. Problem poligraphii i izdat. dela* **2** 44–54
[9] Zuravleva G N, Komarova L Y and Kondratov A P 2015 *Izv.Vys.Uchebn. Zaved. Problem Poligraphii i izdat. dela* **1** 20–6
[10] Syrbu S A, Salikhova A H and Fedorinov A S 2014 *Technosphere safety Technologies* **5** http://ipb.mos.ru/ttb/2014-5/2014-5.html
[11] Kondratov A P, Nagornova I V and Varepo L G 2019 *Journal of Physics: Conference Series* **120** 012067
[12] Syrbu S A, Salikhova A H and Fedorinov A S *Pat.* 128235 RF *on application no.* 2014128235 of 09.07.2014 D06B1/00 flame Retardant for special purpose fabrics of moleskin type based on Aflamite-SAP and silicon dioxide

Comparison of Economic and Transport Capabilities of Heavy C-type Airfield-Based WIG [Wing-in-Ground-Effect] Craft Versus Passenger Aircraft

A N Luchkov[1]

[1] Moscow Aviation Institute (National Research University), Moscow, Russia

E-mail: a.luchok.n@gmail.com

Received xxxxxx
Accepted for publication xxxxxx
Published xxxxxx

Abstract

This article assesses the potential increase in the transport and economic characteristics of heavy cargo and passenger aircraft, if they are processed into WIG craft. This type of air transport is of interest from the point of view of improvement of characteristics from the position of mass use on the airways that pass mainly over the water surface of seas and oceans, as 75% of international air transport is accounted for by these airways. Classic amphibious WIG crafts are known to have extremely low performance due to their high aerodynamic resistance, high structural weight and increased fuel consumption in take-off/landing modes. The solution to this problem may be to switch to airfield take-off WIG crafts. As part of the work, boundary conditions for the study were developed, and a set of formulas was developed to allow for a recalculation of fuel and payload redistribution depending on the operating conditions. The results include changes in gross weight, transport and fuel efficiency, as well as potential changes in aircraft operating costs.

Keywords: WIG Craft, efficiency, airfield takeoff, aerodynamic quality, fuel consumption

1. Introduction

Currently, there is a problem of increasing the speed of cargo delivery to remote areas, separated by large water areas. One of the solutions to this problem may be the use of WIG [wing-in-ground-effect] craft, which, as shown in [1], can provide much higher speeds of movement than river and sea vessels, with higher aerodynamic and seaworthiness (including amphibiousness) compared to hydroplanes.

The work on the study of various layouts and schemes of heavy transport screens was carried out, for example, in the corporation Boeing. As part of the Pelican Ultra program, a super WIG craft project with a maximum takeoff weight of 2700 and 5000 tons, a payload of 1280 and 2500 tons, respectively, and a cruising speed of 490 km/h was piloted and developed. In the course of analysis of transport scheme variants, the drawbacks of marine WIG craft of the aircraft scheme were revealed. Thus, due to the peculiarities of take-off/landing operations from the water there are increased requirements to their strength. Typical requirements for them are wing design with low values of relative elongation, horizontal stabilizer of large area and range, excessively powerful engines to ensure effective takeoff from the water surface. All these negative factors significantly reduce the practical value of the marine WIG craft and make it ineffective in flight. As a result of the research, Boeing specialists decided to design the Lands' without presenting water-based conditions. This resulted in a decrease in the weight of the empty WIG craft by 20–25% due to a decrease in the requirements for the frame strength, which led to a decrease in the weight of the vehicle and the required power plant capacity.

Economically, the effect of the screen reduces the fuel consumption of a cruise WIG plane, which is not directly related to its basing system, which is important for the operation of a high-speed ship. Currently, there are two concepts of using WIG-plans in transportation systems. The first concept — "airport — water area — airport" — provides for takeoff of the WIG-plane from the airfield, flight at an altitude of 6–11 km or over the water area in the near-screen mode and landing at the airport of destination. The second concept — "seaport — water area — seaport" — provides for the basing of the WIG-plane on the coastline, including landing on water, if necessary, and landing in the port of destination [2].

The experience of operation of Russian heavy WIG-plans of the concept "seaport — water area — seaport" with blowing and its subsequent detailed analysis performed in several countries of the world [3–9] has revealed a few their significant drawbacks:

- low aerodynamic quality (within 13–16 units);
- low relative useful mass (useful return);
- significant energy-to-weight ratio (directly for the WIG craft plan, due to the lift and mainframe engines);
- high power consumption on take-off/landing modes;
- insufficient seaworthiness (mainly of light screen-planes);
- low comfort and environmental friendliness.

In order to improve the operational and economic performance of heavy WIG-planes and reduce the financial costs of maintaining the overall life cycle, it is proposed to avoid takeoff/landing from the water surface, which in turn will lead to:

1. Refusal to use the lifting-march propulsion system, which will be replaced (to compensate for the thrust-to-weight ratio at altitude) by two-circuit propulsion systems with a high degree of dual-circuit efficiency, and consequently, to reduce fuel consumption;
2. Reducing the hull strength requirements of the vessel and, therefore, reducing the weight of the glider structure;
3. Redistribution of useful fuselage volumes;
4. Possibilities to install a large elongation wing with equivalent bearing surface area to increase the aerodynamic quality of K.

2. Assessment of changes in the transport efficiency of passenger and cargo aircraft

The perfection of a transport vessel as a complex technical system is characterized by various types of efficiency (weight, economic, transport, operational, etc.), for which several efficiency criteria have been developed. Each of them has its own rational field of application, objectively characterizing those or other aspects of creation and functioning of a ship and allowing to solve a certain range of tasks of this or that stage of a life cycle of an aircraft (its conceptual or general designing, manufacturing, operation, etc.).

We will accept as criteria of mass and transport perfection of cargo and passenger WIG-plane the relation of mass of the empty equipped WIG-plane to ton-kilometer of transported cargo:

$$\mathrm{A} = \frac{m_{oe}}{(m_p * L)}, \qquad (1)$$

where: m_{oe}— operational empty weight, t; m_p— payload, t; L — flying range, km.

With this methodical approach, the mass of an empty WIG-plane is no less important than the weighted return on the WIG-plane's transport efficiency. At the given values of payload and flight range the most economical WIG craft plane diaphragm with a minimum value of m_{oe}. The size and, consequently, tonnage of the WIG-plane determine its load-carrying capacity and flight range, therefore, it is expedient to refer the mass of the empty loaded plane to the product of the flight range and payload m_p *L.

The first part of the study assumes calculation of transport efficiency for passenger and cargo category aircraft, the second part — for WIG-planes.

3. Boundary conditions

The following assumptions and restrictions were made in order to study changes in the transport efficiency of heavy passenger and cargo aircraft:

The study was carried out for two aircraft loading cases:

a. Maximum take-off weight (maximum payload + fuel reserve, $L_p = const$);

b. Maximum take-off weight (maximum fuel volume + payload, $L_p = const$);

In the calculations of transport efficiency, it was assumed that the WIG crafts have been designed by deep processing of the aircrafts studied;

The maximum take-off weight, the relative weight of the empty equipped basic version of the aircraft and the screen-plane created on its basis are equal;

Maximum fuel volumes of the basic versions of aircraft and WIG-planes are equal;

Fuel costs for take-off of basic versions and WIG-plans are equal;

Cruising speeds, aerodynamic quality, and consumable characteristics of passenger and cargo aircraft were taken from open sources [10–13].

Cruising speed for WIG versions of aircraft was assumed to be 450 km/h;

Average flight aerodynamic quality of K was assumed to be 40 (flight support at short and ultra-short distances from the screen);

Specific fuel consumption of Ce (H=0; M=0) and throttle characteristic of φ for airplanes and WIG-planes are assumed to be equal;

High-speed characteristics of power units were taken from open sources;

In order to get the maximum values of transport efficiency of the WIG -plans it was assumed that the whole flight at a given distance passes near the screen;

The study of transport efficiency of passenger, cargo planes and WIG-planes was carried out from the condition that $L_p = const$ in each calculated case;

It is assumed that during the flight on the screen, considering the accepted weight, aerodynamic and flight characteristics of the aircraft, will have the best indicators of kilometer and hourly fuel consumption. Therefore, based on condition 12, the released mass of fuel should be compensated for by the payload to meet condition 1.

The transport efficiency was calculated for the following aircraft groups:

a. Passenger planes: A310-200, A310-300, A318-100, A319-100, A320-200, A321-100, A321-200, A319neo, A320neo, A321neo, A330-200, A330-300, A340-200, A340-300, A340-500, A340-600, A350-800, A350-900, A350-1000, A380, MD-83-88, MD-90, B737-700, B737-800, B737-900, B747-100, B747SP, B747-200, B747-300, B747-400, B747-8I, B757-200, B757-300, B767-200, B767-300, B767-300, B767-30ER, B767-400, B777-200, B777-200ER, B777-200LR, B777-300, B777-300ER, B787-9, Tu-204-120, Tu-214, Tu-204-300, Il-96-300, Il-96-400M, MC-21-300. The graphs below show the division into medium and long-haul by maximum takeoff weight; all aircraft belong to the same category in accordance with ICAO standards.

b. Cargo aircraft: Tu-204, B727, B737-700, B757-200, DC-10, A300, Tristar, Il-96-400T, B747-400F, B747-8F, B767-200, A380-800F, An-124, Il-76T, An-12, L-100, L-500, An-225.

4. Analysis of changes in the transport efficiency of passenger and cargo aircraft

Calculation case 1. Calculation of transport efficiency was performed by formula (1). To calculate the transport efficiency of cargo and passenger screens, the formula (2) modified using formulas from sources [10, 14] was used, considering changes in flight, aerodynamic and consumable characteristics of aircraft.

$$A = \frac{m_{oe}}{(m_p + (m_f - \left(1 - \frac{1}{EXP\left(\frac{LCe_{WIG}}{K_{WIG}V_{WIG_c}}\right)}\right) * m_0) * L_{tow}} \quad (2)$$

where: m_0 — maximum takeoff weight, t; m_f — weight of fuel, t; m_p' — payload in the WIG-plan mode, t; L_p — practical range, km; m_p — payload, t; Ce_{WIG} — WIG specific fuel consumption; K_{WIG} — WIG aerodynamic quality; V_{WIG} — WIG cruising speed.

Calculation case 2. The payload for Calculation Case 2 can be determined as the difference between the maximum takeoff weight m$_0$ and the weight of the loaded and fully loaded aircraft. The following formula was used to calculate the transport efficiency of aircraft:

$$A = \frac{m_{oe}}{(m_0 - m_f - m_{oe}) * \left(\frac{K_a V_a}{Ce_a} Ln\left(\frac{1}{1 - \overline{m}_f}\right)\right)}, \quad (3)$$

where: m_p — payload in the WIG-plan mode, t; $\overline{m}_f$ — fuel fraction, t; m_p— payload, t; Ce_a — aircraft specific fuel consumption; K_a — aircraft aerodynamic quality; V_a — aircraft cruising speed.

It is proposed to use formula (4) for WIG-planes, which is a modification of formula (3) by recalculation of payload mass due to reduction of fuel mass during movement in WIG-plan mode.

$$A = \frac{m_{oe}}{(m_p + (m_f - \left(1 - \frac{1}{EXP\left(\frac{LCe_c}{K_{WIG}V_{WIG}}\right)}\right) * m_0) * \left(\frac{K_a V_a}{Ce_a} Ln\left(\frac{1}{1 - \overline{m}_f}\right)\right))} \quad (4)$$

For Case 2, the average increase in transport efficiency is 2-fold. In both cases, the increase in transport efficiency is due to an increase in mass output.

5. Analysis of changes in economic efficiency of passenger and cargo aircraft

Economic efficiency (EE) is an indicator that demonstrates the cost of aircraft operation, income and net profit. When calculating, the dimension of EE, as a rule, is taken as an equivalent for passenger planes [rub (USD)/1 pass.], for cargo — [USD(USD)/1 kg of cargo]. EE is one of the most important indicators of aircraft competitiveness along with transport efficiency. One of the indicators of EE is the cost price, which consists of many difficult to calculate parameters. One of the most significant cost parameters is the cost of fuel and lubricants, which according to ICAO documentation is about 33% [9]. Within the framework of this work, an assessment of the change in fuel efficiency when comparing aircraft and WIG-planned vehicles was made.

Fuel efficiency. Fuel efficiency, i.e. fuel consumption per unit of transport work, is one of the most important parameters for evaluating the efficiency of transport category aircraft. The level of fuel efficiency mainly depends on specific engine fuel consumption, aerodynamic and weight perfection of the aircraft, its passenger capacity (carrying capacity).

Calculation case 1. The following expressions [4] will be used as the basic calculation formula for determining fuel efficiency:

For passenger aircraft:

$$q = \frac{m_f}{N_{pass} L_p} \left[\frac{g}{pass * km}\right], \quad (5)$$

For cargo aircraft:

$$q=\frac{m_t}{m_p L_p}\left[\frac{g}{kg*km}\right], \qquad (6)$$

where: m_f — weight of fuel, g; N_{pass} — number of passengers; L_p — practical range, km; m_p— payload.

We obtain the mass of fuel for passenger and cargo screen-planes by recalculation according to the Breguet formula [15] with the condition of changing the basic flight-technical, aerodynamic and consumable characteristics in accordance with the conditions. We will accept the following formula to calculate the fuel efficiency of passenger screens:

$$q=\frac{m_0*(1-\frac{1}{EXP(\frac{LCe_c}{K_{WIG}V_{WIG}})})}{(N_{pass}+\frac{(m_f-m'_f*m_0)}{m_{Op\&l}})*L_p}, \qquad (7)$$

For cargo WIG-planes:

$$q=\frac{m_0*(1-\frac{1}{EXP(\frac{LCe_{WIG}}{K_{WIG}V_{WIG}})})}{(m_p+(m_f-m'_f*m_0))*L_p}, \qquad (8)$$

where: m_0 — maximum takeoff weight, kg; m'_f — WIG fuel fraction; $m_{Op\&l}$ — weight of one passenger with luggage, kg; N'_p — number of passengers; L_p — practical range, km; m'_p — payload in the WIG-plan mode, t; Ce_{WIG} — WIG specific fuel consumption; K_{WIG} — WIG aerodynamic quality; V_{WIG} — WIG cruising speed.

Calculation case 2. Calculation Case 2 involves a recalculation of the fuel efficiency for a fully fueled aircraft with a maximum take-off weight.

The fuel efficiency of passenger aircraft can be calculated using the following formula:

$$q=\frac{m_f}{(\frac{(m_0-m_f-m_{CH})}{m_{Op\&l}})*(\frac{K_aV_a}{Cea}Ln(\frac{1}{1-m_f}))}, \qquad (9)$$

Cargo aircrafts:

$$q=\frac{m_f}{(m_0-m_f-m_{CH})*(\frac{K_aV_a}{Cea}Ln(\frac{1}{1-\overline{m}_f}))}, \qquad (10)$$

where: $\overline{m}_f$— fuel fraction; m'_p — payload in the WIG-plan mode, t; Ce_a — aircraft specific fuel consumption; K_a — aerodynamic quality; V_a — cruising speed.

The fuel efficiency of passenger WIG-planes can be determined by the formula:

$$q=\frac{m_0*(1-\frac{1}{EXP(\frac{LCe_{WIG}}{K_{WIG}V_{WIG}})})}{(N_{pass}+\frac{(m_f-m'_f*m_0)}{m_{Op\&l}})*(\frac{K_aV_a}{Cea}Ln(\frac{1}{1-\overline{m}_f}))}, \qquad (11)$$

Cargo WIG-planes:

$$q=\frac{m_0*(1-\frac{1}{EXP(\frac{LCe_{WIG}}{K_{WIG}V_{WIG}})})}{m_p+(m_f-m'_f*m_0)*(\frac{K_aV_a}{Cea}Ln(\frac{1}{1-\overline{m}_f}))}, \qquad (12)$$

6. Conclusions

Based on the studied literature, it was possible to get acquainted with a complex of advantages and the lacks designed WIG-planes. The main drawbacks of the existing WIG-planes are the relatively high weight of the construction, power system, increased fuel consumption especially in takeoff modes and low aerodynamic quality (13–16 units), which in turn has a negative impact on the efficiency and effectiveness. The easiest way to solve these problems is to switch from the concept of an amphibious WIG-plane to WIG-plane, which can be based only on prepared airfields.

Heavy passenger WIG-planes type C of the "Airport — water area — airport" concept are a prospective type of air transport from the point of view of increasing transport efficiency by increasing the weight of commercial return (up to 70–75% on average), by reducing fuel consumption while moving in the area of the screen effect. At the same time, the best indicators of weight efficiency are those aircrafts, 100% of the way of which passes over the water surface. The general trend of change in the total weight return can be estimated from Fig. 1 a, b.

Fuel efficiency of type C WIG-planes can be increased in average by 1.8–2 times due to the increase of aerodynamic quality and reduction of specific fuel consumption of the power unit, which, in general, reduces the cost of operation of such aircraft by 15–17% on average as compared to conventional aircraft (Fig. 2 a, b).

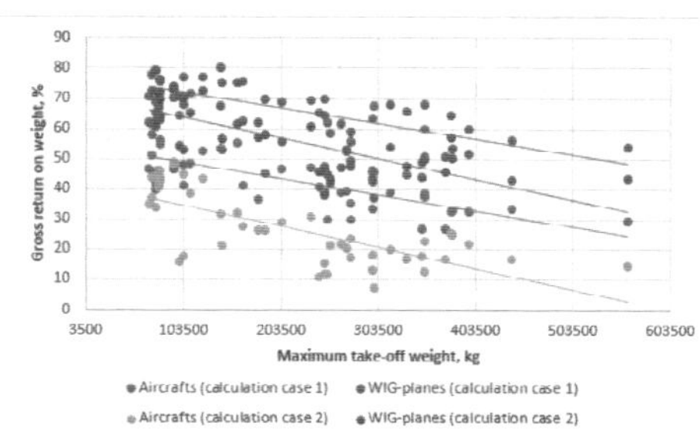

a

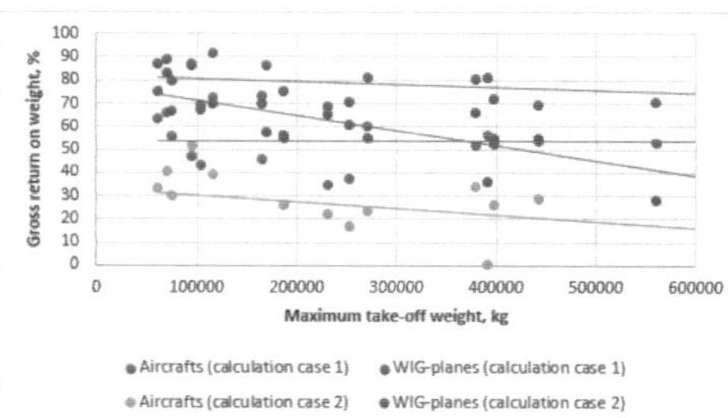

b

Figure 1. Change in the total weight return of passenger (a) and cargo (b) aircraft.

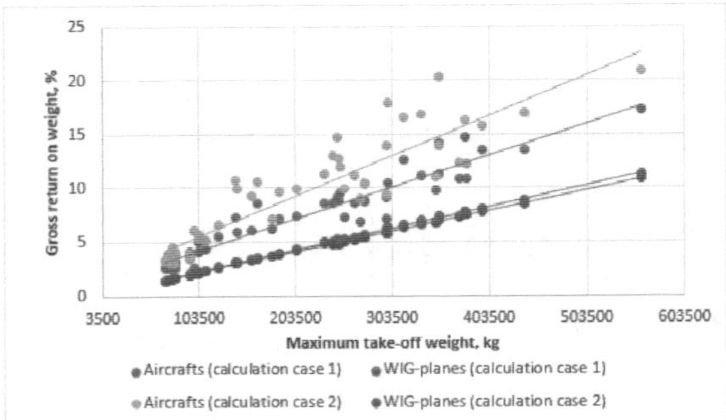

Figure 2. Change in kilometer-long fuel consumption of passenger and cargo aircraft.

References

[1] Yun L, Bliault A and Doo J 2010 *WIG Craft and Ekranoplan* (New York: Springer)

[2] Panchenkov A N, Drachev P T and Lyubimov V I 2006 *Expertise of the screen-planes* (Novgorod: VGAVT)

[3] Jia Q, Yang W and Yang Z 2016 *Numerical study on aerodynamics of banked wing in ground effect* (Shanghai, China: Tongji University)

[4] Jamei S[a], Maimun A and Azwadi N[c] 2018 *Ground boundary layers effect on aerodynamic coefficients of a compound wing with respect to design parameters* (Iran: Department of Marine Engineering, Faculty of Engineering, Persian Gulf University, Boushehr)

[5] Park H 2017 *MATEC Web of Conferences* **139** 00022

[6] Kinaci O K, Cosgun T, Yurtseven A and Vardar N 2017 *Ground effect or grounding: Which happens when?* (Istanbul, Turkey: Yildiz Technical University)

[7] Mascalik A I, Nagapetyan R A and Lukyanov A I 2013 *WIG-planes transport ships of the future* (St. Petersburg: Shipbuilding)

[8] Nebylov A and Nebylov V 2014 Wing-in-ground effect vehicles flight automatic control systems development problems. *5th AEROTECH conference Kuala Lumpur, Malaysia* **629** pp 370–5

[9] Gadalov V V, Gappoev M A and Kuteynikov M A 2007 Wing-in-ground (WIG) craft (ekranoplan). Practical aspects of the classification and survey according to RS instruments *9th International Conference on Fast Sea Transportation Shanghai, China* pp 76–9

[10] Fortinov L G, Kobyzev G, Zabaluyev I and Sokolyansky V 2002 Assessment of the average aerodynamic quality of mainline jets on cruising modes of flight *Proceedings of the conference Gelendzhik-2002* (Moscow: TsAGI printing house) pp 200–4

[11] Fortinov L G 2002 *Synthesis of hydro-aviation aircraft appearance and methodology of their complex assessment at the initial stages of design* (Moscow: Moscow State Aviation Institute)

[12] Arutyunov A G, Dydyshko D V and Kuznetsov K V 2016 History of transport aircraft development *Proceedings of MAI* **89** http://trudymai.ru/published.php?ID=72654

[13] Arutyunov A G, Dydyshko D V, Yendogur A I, Kuznetsov K V and Tolmachev V I 2016 Prospects for the development of transport aircraft *Proceedings of MAI* **90** http://mai.ru//upload/iblock/d01/arutyunov_dydyshko_endogur_kuznetsov_tolmachev_rus2_1.pdf

[14] Shaynin V M 1962 *Weight and transport efficiency of passenger planes* (Moscow: Oboronghiz)

[15] Torenbeek E 1976 *Synthesis of subsonic airplane design* (Rotterdam: Nijgh-Wolters-Noordhoff Universitaire Uitgevers B V)

IOP Conf. Series: Materials Science and Engineering **714** (2020) 012019 doi:10.1088/1757-899X/714/1/012019

Development of Software for Computing of Aircraft's Takeoff and Landing Characteristics Using Expert System Technology

A V Melnichuk[1], V A Nesterov[1] and V A Sudakov[2,3]

[1] Moscow Aviation Institute, Moscow, Russia
[2] Keldysh Institute of Applied Mathematics, Moscow, Russia
[3] Applied Modelling Science Laboratory, Plekhanov Russian University of Economics, Moscow, Russia

E-mail: alexander.melnichyuk@gmail.com

Received xxxxxx
Accepted for publication xxxxxx
Published xxxxxx

Abstract

The principles of expert system are proposed in the development of software application for calculations of rational takeoff and landing characteristics. The modular architecture is developed for the process of takeoff and landing parameters calculations in the system. A prototype of a client-server software application for electronic flight bag based on tablet computer was developed on the example of a specific type of Russian aircraft.

Keywords: take-off and landing characteristics, expert system

1. Introduction

Despite the fact that takeoff and landing are the shortest phases of flight, these are the most difficult and critical in terms of flight safety. At the same time, calculation of takeoff and landing characteristics (which also known as aircraft performance) is of great importance for ensuring safety of these phases. The main purpose of this calculations is to determine maximum allowed takeoff weight of an aircraft, speeds for takeoff (V_1 — the decision speed, i.e. the maximum speed to which takeoff can be safely terminated; V_R — the front landing gear lifting speed; V_2 — the safe takeoff speed) and landing (V_{REF} — the approach speed), maximum landing weight of an aircraft.

Dependencies of flight characteristics are presented in aircraft flight manual (AFM) as nomograms based on mathematical model of a specific aircraft modification and adjusted in accordance with results of flight tests.

The values of takeoff and landing characteristics depend on many operational conditions: the actual takeoff mass of the aircraft, the outside air temperature, the barometric pressure, the characteristics of the runway (declared lengths, slope, condition of its surface), obstacles in the area of the airport, wind's speed and direction. A significant influence is also exerted by the pilot's control actions by adjusting such aircraft parameters as engine thrust, flaps position, etc. Also limitations due to presence of permissible deferred defects and deviations of the aircraft configuration (according to MEL — minimum equipment list and CDL — configuration deviation list) or determined by the operator's policy can be taken into account in calculations.

Currently specialized software applications (such as "Boeing Onboard Performance Tool" for aircrafts manufactured by Boeing Corporation and "FlySmart+" for Airbus aircrafts) become widely used for calculating of takeoff and landing parameters [1, 2]. These applications are intended for use as software of electronic flight bag (EFB) based on tablet computers.

IOP Conf. Series: Materials Science and Engineering **714** (2020) 012019 doi:10.1088/1757-899X/714/1/012019

Using the above software allows to speed up aircraft performance computing process significantly and define the rational values of such parameters as thrust, flaps position which increase the economic efficiency of aircraft flight operations (that results in saving engine resources, reducing fuel consumption, increasing of commercial load) while ensuring a high level of takeoff and landing safety. Nevertheless, these software solutions don't allow operators to implement its own rules for special limitations in aircraft performance calculations in accordance with its policy.

In turn, currently there's no such software applications for aircrafts produced by russian manufacturers. Pilots of russian aircrafts define takeoff and landing characteristics manually using special nomograms or tables presented in AFM. Performing calculations in accordance with this method is a long process that requires increased attention, and use of incorrect calculation results can lead to an aviation incident or a disaster.

Thus, develop of general principles of the special aircraft performance calculating system for pilots is an actual scientific and technical task.

The scientific novelty of the study is in use of the expert systems technology that provides flexibility in definition of takeoff and landing characteristics by ability of applying special restrictions in the computed results in accordance with the rules based on MEL, CDL or operator's policy.

Creation of the special language for knowledge description is proposed to manage rules in the expert system. This approach will provide operators with the ease of creating rules and filling the knowledge base of an expert system without the need of make changes to the software code.

2. Task definition

The system for aircraft performance calculations must comply with the requirements for EFB software applications presented in ICAO Doc 10020 guidance material [3]. The experience of using such software for calculations of takeoff and landing performance of foreign-made aircrafts indicates a number of errors associated with the entry and interpretation of data [4, 5]. Therefore, in order to reduce the risk of errors, it's necessary to take into account ICAO guidance material principles in designing of system's graphical user interface.

The system being developed consists of blocks is shown in the figure 1.

Dividing of the system into blocks is necessary to ensure its flexibility and extensibility, allowing to take into account the specifics of different types of aircraft.

The filling of the calculation module can be performed either by pre-calculated specific tables of runway analysis for a particular aircraft type, or by electronic data (digitized nomograms) from the flight manual.

The official aeronautical information sources is used to fill the airport characteristics database.

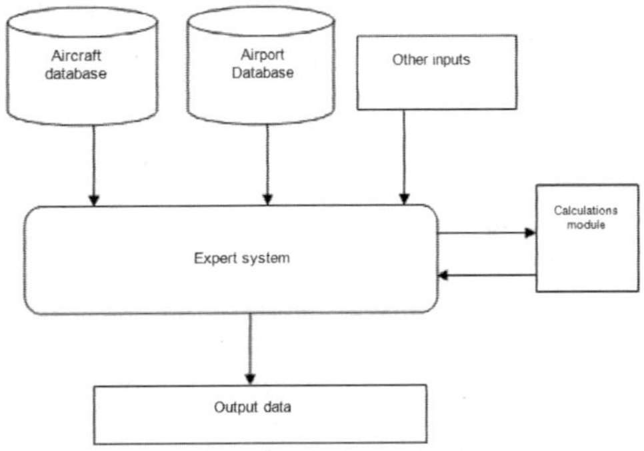

Figure 1. Architecture of a performance calculations process.

A fragment of the database logical structure is shown in the figure 2. The table "Airport data" includes tuples of ICAO airport codes, airport names and their elevations. The "Runway data" table includes tuples related to the characteristics of the runways: their identifiers, declared distances (TORA — Takeoff Run Available — available takeoff distance, TODA — Takeoff Distance Available, ASDA — available Accelerate-Stop Distance Available, LDA — Landing Distance Available) and slopes values. The table "Obstacles data" includes tuples of obstacle identifiers, their types, distance from the end of the runway, the values of lateral offsets and heights.

The specified database is subject to frequent updates. Its relevance, integrity, accuracy and quality are critical in the process of aircraft performance computing.

The knowledge base of the expert system contains rules describing the limitations imposed by the operator and affecting take-off and landing performance.

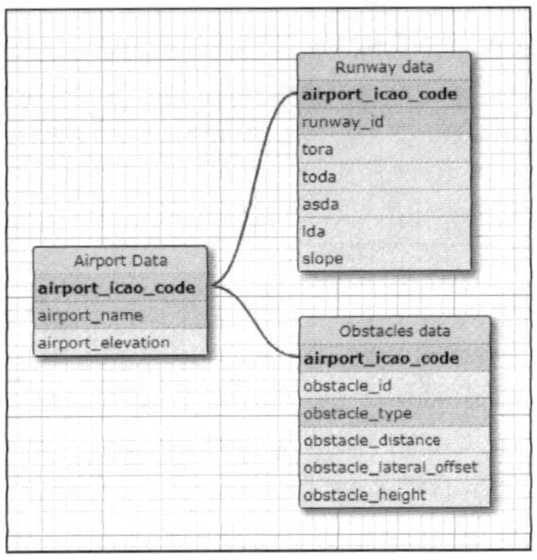

Figure 2. Fragment of the structure of airport characteristics database.

3. EFB platform selecting task

According to ICAO Doc 10020 [3], EFB hardware is divided into 2 classes: portable and installed (i.e. integrated into the aircraft). Hardware of portable EFB is an electronic device such as a tablet computer.

Tablet computers become widely used in aviation as portable EFB devices [6, 7] because of their lower cost (in comparison of installed EFB), the technology of operations with portable EFB is not limited to use only in the cockpit, and this equipment is approved by aviation authorities. Currently market presents a large number of different models of tablet computers that can be used as EFB. With the development of this market and the regulatory framework governing the approval and use of tablet computers in aviation as an EFB, aircraft operators are faced with the challenge of choosing the most rational device model for them.

The task of selecting a rational model of tablet computer was considered in [8].

Taking into account the existing variety of electronic tablets, and to ensure the operation of the software on the domestic element base to minimize dependence on foreign manufacturers, the main goal is to create a universal system for aircraft performance calculating without referencing to a specific hardware platform.

The achievement of this goal is possible both by adapting the software-algorithmic prototype of the system to existing platforms, which is a simple technical task, and by developing using universal programming systems that translate the code for the corresponding virtual machines, that provides the ability to use the system on the whole variety of existing devices, without need of developing individual version for each of them. Currently such universal solutions are React Native and Flutter programming systems. Despite the fact that these systems are developed mainly by foreign developers, they are free and distributed with open source code, that reduces the risks of termination of support and allows to independently modify the system libraries and minimizes the likelihood of malicious "bookmarks" in the code.

4. The software prototyping

As the first stage of the software application developing is the task of creating a software-algorithmic prototype, the choice of the initial hardware platform and operation system is not critical. Therefore, in order to simplify the testing of the developed prototype and for possibility of rapid adaptation to other platforms due to the simplicity of the language syntax, its development is carried out in the Xcode programming environment with use of Swift language.

Figure 3 shows the interface of the software prototype.

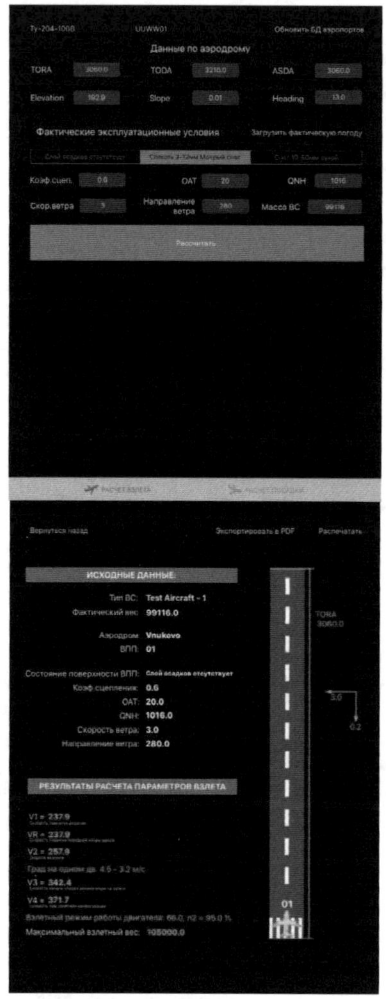

Figure 3. The interface of the software prototype.

The algorithm of aircraft takeoff performance calculations is built on performance data in the aircraft flight manual and consists of two stages: determining of the maximum take-off weight and determining the characteristic speeds.

To automate the calculation, nomograms were digitized and their mathematical models were built. For example, in accordance with the constructed mathematical model, the takeoff mass limited by the normalized climb gradient is presented as the function:

$$m_{TO} = f(T, H_{aer}), \tag{1}$$

where T — air temperature at the aerodrome, H_{aer} — the aerodrome height that corresponding to the actual barometric pressure at the aerodrome.

Depending on the values that the parameters T and H_{aer} takes, the take-off mass limited by the climb can be determined in 12 different ways. For example, if $15 \le T \le 20$ and $1800 \le H_{aer} \le 2000$, m_{TO} is determined in accordance with the formula:

$$m_{TO} = (4 - 0.2T)[(118061 - 8.48485H_{aer}) - -(119430 - 10.596H_{aer})] + (119430 - 10.596H_{aer}). \quad (2)$$

The algorithm of landing performance calculations also consists of two stages: determining of the maximum landing weight and the approach speed V_{REF}.

For the database of airport characteristics, information from AIP (Aeronautical Information Publications) is used. AIP is a publication issued by or with the authority of a state and containing aeronautical information essential to air navigation.

5. Conclusion

The use of expert system technology in the developed software provides flexibility in determining takeoff and landing characteristics and simplifies the process of filling the knowledge base with rules without need of making changes in the program code.

The practical utility of the proposed system is in improve of efficiency and safety of flight operations of aircrafts in large-scale and increase the competitiveness of these by providing the proposed software as a related service that improves the flight operations technology.

Acknowledgments

The reported study was funded by RFBR according to the research project № 18-00-00012 (18-00-00011).

References

[1] Babb T A and Hiers C 2018 *International Journal of Aviation, Aeronautics, and Aerospace* **5** 1–101
[2] Tump R S, G W H van Es, Karwal A K and Verbeek M J 2015 Best practices for approval of Performance and MB applications on EFBs (EASA Report)
[3] ICAO Doc 10020 Manual of electronic Flight Bags (EFB) 2016
[4] Chandra D C & Mangold S J 2000 *Proceedings of the 19th Digital Avionics Systems Conference* **2** 5A1/1–5A1/7
[5] Divya C 2002 *Proceedings of the HCI-02 Conference* pp 69–73
[6] Haddock K and Beckman W 2015 *Collegiate Aviation Review International* **33** 1–13
[7] Skaves P 2011 *Digital Avionics Systems Conference (DASC)* pp 8D1–1
[8] Sypalo K, Nesterov V, Dutov A and Sudakov V 2018 *Journal of Computer and Systems Sciences International* **2** 230–8

IOP Conf. Series: Materials Science and Engineering **714** (2020) 012020 doi:10.1088/1757-899X/714/1/012020

Design of Cylindrical Metal Panels with Mild Camber Based on Post-Buckling State when Compressed Taking into Account Initial Deflection

O V Mitrofanov[1]

[1] Moscow Aviation Institute (National Research University), Moscow, Russia

E-mail: Oleg1Mitrofanov@yandex.ru

Received xxxxxx
Accepted for publication xxxxxx
Published xxxxxx

Abstract

Applied design method for cylindrical metal panels with mild camber with possible post-buckling state when compressed is suggested. Method is based on analytical solution of geometrically non-linear task and is reduced to non-linear equation solution in panel thickness when reaching critical stress with post-buckling behaviour.

Keywords: cylindrical panel, post-buckling behavior, initial deflection, compression

1. Introduction

Let us consider task of the best possible (reasonable) supersonic aircraft fuselage section smooth thin skins thickness definition. In this case we shall assume that panels with mild camber are loaded mainly in compression and have initial deflection due to taken structural and technological schemes with welding used. This task numerical solution with possible post-buckling behavior and under critical loading using geometrically non-linearity is examined in this investigation [1]. Also, investigations [2–4] should be mentioned because research results for coupled thermomechanical problems using finite element method are shown in them.

In this paper we shall examine analytical solution of design problem using geometrically non-linear correlations taking into account initial deflection of thin isotropic cylindrical panels with mild camber and design methodology for load-bearing composite panels based on post-buckling state [6].

Let us write main geometrically non-linear correlations for thin panels. Equation of strain compatibility taking into account initial deflection (ID) if of following form [6]

$$L_1(F) - L_2(W) = 0, \qquad (1)$$

with $L_1(F) = \frac{1}{E\delta}\left[\frac{\partial^4 F}{\partial x^4} + 2\frac{\partial^4 F}{\partial x^2 \partial y^2} + \frac{\partial^4 F}{\partial y^4}\right]$,

$$L_2(W) = \left(\frac{\partial^2 (W + W_{id})}{\partial x \partial y}\right)^2 - \left(\frac{\partial^2 W_{id}}{\partial x \partial y}\right)^2 - \frac{\partial^2 (W + W_{id})}{\partial x^2}$$

$$\frac{\partial^2 (W + W_{id})}{\partial y^2} + + \frac{\partial^2 W_{id}}{\partial x^2}\frac{\partial^2 W_{id}}{\partial y^2} - \frac{1}{R}\frac{\partial^2 W_{id}}{\partial x^2}.$$

Non-linear equation of Karman type for isotropic panels taking into account ID

$$L_3(F,W) - L_4(W) = 0, \qquad (2)$$

with $L_4(W) = D\left[\frac{\partial^4 W}{\partial x^4} + 2\frac{\partial^4 W}{\partial x^2 \partial y^2} + \frac{\partial^4 W}{\partial y^4}\right]$,

$$L_3(F,W) = \frac{\partial^2 F}{\partial y^2}\frac{\partial^2 (W + W_{id})}{\partial x^2} + \frac{\partial^2 F}{\partial x^2}\frac{\partial^2 (W + W_{id})}{\partial y^2} -$$

$$-2\frac{\partial^2 F}{\partial x \partial y}\frac{\partial^2 (W + W_{id})}{\partial x \partial y} + \frac{1}{R}\frac{\partial^2 F}{\partial x^2}.$$

With $D(\delta)$ being skin bending stiffness (thickness), F being Airy function for which following correlations for membrane stress are valid

$$\sigma_x = \frac{1}{\delta}\frac{\partial^2 F}{\partial y^2},\ \sigma_y = \frac{1}{\delta}\frac{\partial^2 F}{\partial x^2},\ \tau_{xy} = -\frac{1}{\delta}\frac{\partial^2 F}{\partial x \partial y}. \tag{3}$$

Equations (1) and (2) are initial system of geometrically non-linear equations for post-buckling behavior research of isotropic panels with ID.

2. Geometrically non-linear tasks of confirmatory analysis

Investigation [1] shows examples of welding use along the longitudinal skin edge with ID origination and following waves values: in longitudinal direction $m=1$ is derived and in transverse direction $n=2$ or $n=3$ are derived. Further on when considering analytical solutions of geometrically non-linear tasks we shall use the same examples. At first some solutions of confirmatory analysis simplified tasks will be shown and then we shall represent panels design method variant based on post-buckling state taking into account initial deflection.

a) Hinge-supported panel with initial deflection with $m=1,\ n=2$.

Let us examine at first hinge-supported cylindrical panel with mild camber and IF which can be approximated by following function

$$W_{id} = f_{id}\cdot \sin\frac{\pi x}{a}\sin\frac{2\pi y}{b}. \tag{4}$$

Now let us represent deflection in similarly

$$W = f\cdot \sin\frac{\pi x}{a}\sin\frac{2\pi y}{b}. \tag{5}$$

After substitution of equations (4)–(5) into non-linear equation of strain compatibility (1) following equation is derived

$$L_1(F) = \frac{2\pi^2\left(f^2+2ff_{id}\right)}{a^2b^2}\cdot\left(\cos\frac{2\pi x}{a} + \cos\frac{4\pi y}{b}\right) + \frac{\pi^2 f}{Ra^2}\sin\frac{\pi x}{a}\sin\frac{2\pi y}{b}. \tag{6}$$

Of which stress function can be obtained

$$F = A_1 E\delta\left(f^2+2ff_{id}\right)\cos\frac{2\pi x}{a} + A_2 E\delta\left(f^2+2ff_{id}\right)\cos\frac{4\pi y}{b} + A_3 E\delta\cdot f\cdot\sin\frac{\pi x}{a}\sin\frac{2\pi y}{b} - \frac{T_x y}{2}, \tag{7}$$

With T_x being compression force,

$$A_1 = \frac{a^2}{16b^2\pi^2},\ A_2 = \frac{b^2}{128a^2\pi^2},\ A_3 = \frac{a^2b^4}{\pi^2 R\left(4a^2+b^2\right)^2}.$$

Using Bubnov-Galerkin method procedure following non-linear equation is obtained

$$D\frac{\pi^2 f}{4a^3b^3}\left(4a^2+b^2\right)^2 + E\delta\frac{2\pi^4}{ab}\left(f+f_{id}\right)\cdot\left(f^2+2ff_{id}\right)\left(A_1+A_2\right) + E\delta\frac{\pi^2 bA_3 f}{4Ra} = \frac{\pi^2 bT_x\left(f+f_{id}\right)}{4a}. \tag{8}$$

Which correlates acting compression force, panel thickness δ, flexure amplitude f and ID. So, with given compression force T_x and known panel thickness flexure amplitude f is defined by equation (8) and membrane stress is calculated using Airy stress function (3).

b) Hinge-supported panel with initial deflection with $m=1,\ n=3$

Now let us examine hinge-supported panel with ID which can be approximated by following function

$$W_{id} = f_{id}\cdot\sin\frac{\pi x}{a}\sin\frac{3\pi y}{b}. \tag{9}$$

Let us represent deflection as

$$W = f\cdot\sin\frac{\pi x}{a}\sin\frac{3\pi y}{b}. \tag{10}$$

After substitution of correlations (9)–(10) into equation (1) following equation is obtained

$$L_1(F) = \frac{9\pi^2\left(f^2+2ff_{id}\right)}{2a^2b^2}\cdot\left(\cos\frac{2\pi x}{a}+\cos\frac{6\pi y}{b}\right) + \frac{\pi^2 f}{Ra^2}\sin\frac{\pi x}{a}\sin\frac{3\pi y}{b}. \tag{11}$$

Of which following stress function is obtained

$$F = A_1 E\delta\left(f^2+2ff_{id}\right)\cos\frac{2\pi x}{a} + A_2 E\delta\left(f^2+2ff_{id}\right)\cos\frac{6\pi y}{b} + A_3 E\delta\cdot f\cdot\sin\frac{\pi x}{a}\sin\frac{3\pi y}{b} - \frac{T_x y^2}{2},$$

with $A_1 = \frac{9a^2}{16b^2\pi^2},\ A_2 = \frac{9b^2}{2592a^2\pi^2},\ A_3 = \frac{a^2b^4}{\pi^2 R\left(9a^2+b^2\right)^2}.$

Using Bubnov-Galerkin method procedure for this case following equation is obtained

$$D\frac{\pi^2 f}{4a^3b^3}\left(9a^2+b^2\right) + E\delta\frac{\pi^2}{6ab}\left(f+f_{id}\right)\cdot\left[27\pi^2\left(A_1+A_2\right)\left(f^2+2ff_{id}\right) - 64A_3 f\right] + E\delta\frac{8\pi^2 A_3 f\left(f+f_{id}\right)}{3ab} + E\delta\frac{9\pi^2 A_3 f - 64A_1\left(f^2+2ff_{id}\right)}{36Ra}b = \frac{\pi^2 bT_x\left(f+f_{id}\right)}{4a}. \tag{12}$$

It should be noted that confirmatory analysis procedure is described above but structure of equations (8) and (12) is slightly different.

3. Engineer best possible thickness definition methods for cylindrical panel with mild camber based on post-buckling state

a) Let us consider at first cylindrical panels design task based on post-buckling state assuming that used materials design characteristics are known, there is no ID ($f_{id}=0$) and let us lay down engineer method provisions for fuselage sections load-bearing skins design.

At first let us assume that at limit loading level buckling is unacceptable. In this case thin skins thickness δ_{stab} can be defined based on stability conditions.

Secondly with loads above limit level buckling is acceptable and skin thickness $\delta_{post-buckl.}$ can be defined when reaching critical stress with post-buckling behavior which is described by geometrically non-linear correlations. General method for load-bearing panels design based on post-buckling state is shown in investigation [4].

Thirdly panel minimal thickness $m(\delta_{stab.},\delta_{post-buckl.})$ is defined because generally above-mentioned values are defined at different loading levels using different criteria.

Let us describe method use for hinge-supported panel with $m=1$ and $n=2$ in more details. At first let us transform equation (8) for linear case in order to consider stability task with mild camber ($f\to 0$).

$$\frac{E\delta^3}{12\left(1-\mu^2\right)}\frac{1}{a^3b^3}\left(4a^2+b^2\right)^2+E\delta\frac{bA_3}{Ra}=\frac{bT_x^{\lim}}{a}. \quad (13)$$

That is we have analytical equation for panel thickness determination based on stability conditions at limit loading level. Secondly we shall also transform equation (8) into

$$\begin{gathered}\frac{E\delta^3}{12\left(1-\mu^2\right)}\frac{\pi^2}{4a^3b^3}\left(4a^2+b^2\right)^2+E\delta\frac{2\pi^4}{ab}f^2\left(A_1+A_2\right)+\\+E\delta\frac{\pi^2bA_3}{4Ra}=\frac{\pi^2bT^{ult}}{4a}\end{gathered} \quad (14)$$

with ultimate loads acting. In this case according to method [6] we shall for example assume that with post-buckling behavior longitudinal stress reaches critical static stress $\sigma_x=\sigma_в$. Membrane stress defined by correlations (3) shall be written as second-degree equation in flexure amplitude f if positive stress is compression one

$$\begin{gathered}\sigma_x=\frac{1}{\delta}\frac{\partial^2 F}{\partial y^2}=\frac{1}{\delta}\left[A_2f^2\frac{16\pi^2}{b^2}\cos\frac{4\pi y}{b}+\right.\\\left.+A_3f\frac{4\pi^2}{b^2}\sin\frac{\pi x}{a}\sin\frac{2\pi y}{b}-T_x^{ult}\right]=\sigma_в.\end{gathered} \quad (15)$$

Expressing amplitude f from the last equation and substituting it into equation (14) non-linear equation in required panel thickness with compression flow T_x^{ult} acting can be derived. It should be noted that stress reaches limit values in points defined when considering value

$$\begin{gathered}\Delta=\frac{1}{8a^2}f^2\cos\frac{4\pi y}{b}+\\+\frac{4a^2b^2}{R\left(9a^2+b^2\right)^2}f\sin\frac{\pi x}{a}\sin\frac{2\pi y}{b}.\end{gathered} \quad (16)$$

In this case to the first approximation for cylindrical panels with mild camber ($R\to\infty$) it can be assumed that $y=0, x=a/2$ and further on we shall examine non-linear equation in best possible thickness with stress $\sigma_x=\sigma_в$ in above mentioned points.

b) Now let us consider peculiarities of panels design task based on post-buckling behavior with ID origination ($f_{id}\neq 0$) and with taken technological solutions. Let us assume that ID amplitude f_{id} can be defined experimentally as a dependence on panel thickness

$$f_{id}(\delta)=\Omega(\delta)=A\delta^2+B\delta+C, \quad (17)$$

with A, B, C being coefficients obtained experimentally.

Then substituting ID as function $f_{id}(\delta)$ into correlations of shown above method non-linear equation in panel thickness with acceptability of post-buckling behavior at limit stress and with ID can be obtained. For considered above case we shall transform equation (8) into following form

$$\begin{gathered}E\delta\frac{2\pi^4}{ab}\left(f+A\delta^2+B\delta+C\right)\cdot\\\cdot\left(f^2+2f\left(A\delta^2+B\delta+C\right)\right)\cdot\left(A_1+A_2\right)+\\+\frac{E\delta^3}{12(1-\mu^2)}\frac{\pi^2f}{4a^3b^3}(4a^2+b^2)^2+\\+E\delta\frac{\pi^2bA_3f}{4Ra}=\frac{\pi^2bT_x\left(f+A\delta^2+B\delta+C\right)}{4a}.\end{gathered} \quad (18)$$

and equation (15) for longitudinal membrane stress is of following form

$$\begin{gathered}\sigma_x=\frac{1}{\delta}\frac{\partial^2 F}{\partial y^2}=\frac{1}{\delta}\left[A_2\left(f^2+2f\left(A\delta^2+B\delta+C\right)\right)\cdot\right.\\\left.\cdot\frac{16\pi^2}{b^2}\cos\frac{4\pi y}{b}+A_3f\frac{4\pi^2}{b^2}\sin\frac{\pi x}{a}\sin\frac{2\pi y}{b}-T_x^{ult}\right]=\sigma_в.\end{gathered}$$

Thus, method of item 2(a) remains but equation for panel thickness determination which can be solved only numerically is complicated.

References

[1] Merkulov I and Endogur A 2017 Supersonic aircraft welded section model construction taking into account structural and technological scheme *MAI writings* **94** 30

[2] Murakawa H, Deng D, Ma N and Wang J 2012 *Computational Materials Science* **51** 43–52

[3] Islam M, Buijk A, Rais-Rohani M and Motoyama K 2014 *Finite Elements in Analysis and Design* **84** 54–64

[4] Voutchkov I, Keane A J, Bhaskar A and Olsen T M 2005 *Computer Methods in Applied Mechanics and Engineering* **194** 3535–51

[5] Volmir A 1956 *Flexible plates and shells* (Moscow: Gostechizdat)

[6] Mitrofanov O 2003 *Engineer design methods for load-bearing composite panels* (Moscow: Sputnik+ company

Determination of Aerodynamic Parameters in Various Conditions of Input Signal

S V Nikolaev[1], Moung Htang Om[1], Kyaw Zin Latt[1]

[1] Moscow Aviation Institute, Moscow, Russia

E-mail: moughtangom50@gmail.com

Received xxxxxx
Accepted for publication xxxxxx
Published xxxxxx

Abstract

In this paper, determination of aerodynamic parameters of the aircraft in various conditions of input signal which loses its signal after some times is discussed. Normally distributed random variables are utilized as the measurement noise. The results, which show the accuracy of estimation depending on the condition of input signal, are presented.

Keywords: aerodynamic parameters, input signal, random errors, sliding window, accuracy

1. Introduction

Aerodynamic parameters estimation is widely utilized for monitoring the technical state of the aircraft and the on-board systems [1, 2]. This technology is used in flight dynamic model modifying, flight control law design, flight envelope expansion, flight simulator development, and so on [1]. A sliding windows model is an important case of the streaming model, where only the most "recent" elements remain active and the rest are discarded. The concept of sliding windows expresses the importance of recent data for applications. In this work, normally distributed discrete random variable with zero mean and standard deviation (0.2) is used as measurement noises.

2. Input signal formed by pilot's control

Various types of input signal (φ_t) are formed by the pilot's control according to different conditions. According to some researches [3, 7], it's found that the harmonic input signal assures the better accuracy of the estimation by Least Square Method (LSM). Therefore, in this works, the combination of two sin waves with two different frequencies, which is supposed to be loaded in the elevator module is used as input signal (φ_t). It's assumed that, this input signal loses its signal after 24s of the processing time and it is shown in Fig. 1.

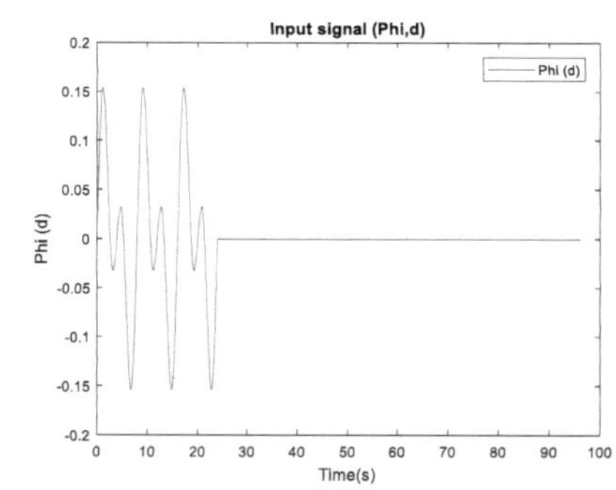

Figure 1. Input signal (φ_t) are formed by the pilot's control.

3. Simulation of the aircraft motion and measurement of input/output signals

Simulation of the aircraft motion must be performed after collecting all the input data. For simulation, it is necessary to form the object model. For the formation of object model, the angle of attack and angular velocity of pitch of the aircraft in discrete form are used.

$$\alpha'(t) = -\bar{Y}^{\alpha}\alpha(t) + \omega_z(t) - \bar{Y}^{\varphi}\varphi(t), \qquad (1)$$

$$\omega_z'(t) = \bar{M}_z^{\alpha}\alpha(t) + \bar{M}_z^{\omega_z}\omega_z(t) + \bar{M}_z^{\varphi}\varphi(t). \quad (2)$$

It is more comfortable to do the identification in discrete form and the mathematical formulas in discrete form for the angle of attack (1) and angular velocity of pitch (2) and they are written as follow;

$$\alpha(t_{i+1}) = \alpha(t_i) + \Delta t[-Y^{\alpha}\alpha(t_i) + \omega_Z(t_i) - \varphi(t_i)], \quad (3)$$

$$\omega_Z(t_{i+1}) = \omega_Z(t_i) + \Delta t[M_Z^{\alpha}\alpha(t_i) + M_Z^{\omega_z}\omega_Z(t_i) - -M_Z^{\varphi}\varphi(t_i)]. \quad (4)$$

Where, $\alpha(t_{i+1})$ — angle of attack for time instant $(t_{(i+1)})$ (radian); $\omega_z(t_{i+1})$ — angular velocity of pitch for time instant $(t_{(i+1)})$ (radian/s); $\alpha(t_i)$ — angle of attack for time instant(t_i); $\omega_z(t_i)$ — angular velocity of pitch for time instant(t_i); $\Delta t = t_{i+1} - t_i$ — time discretization interval; Y^{α}, Y^{φ}, M_z^{α}, $M_z^{\omega_z}$, M_z^{φ} — aerodynamic coefficients to be identified.

For simulation, it was assumed that the registration frequency (*f*-registration) is 32 Hz. This registration frequency (*f*-registration) also can influence the process of estimation and the accuracy of the estimated parameters. In this work, the time sampling interval is $\Delta t = 1/32$.

Firstly, the input signals (φ_t) in this work, angle of attack (3), and the angular velocity of pitch (4) are simulated without any measurement noises. After simulation process, the simulated signals without measurement noises are obtained and shown in Fig. 2. Only after simulation process, the measurement for all input and output signals must be performed. In measurement, the input signal, angle attack (3) and angular velocity of pitch (4) with measurement noises is performed.

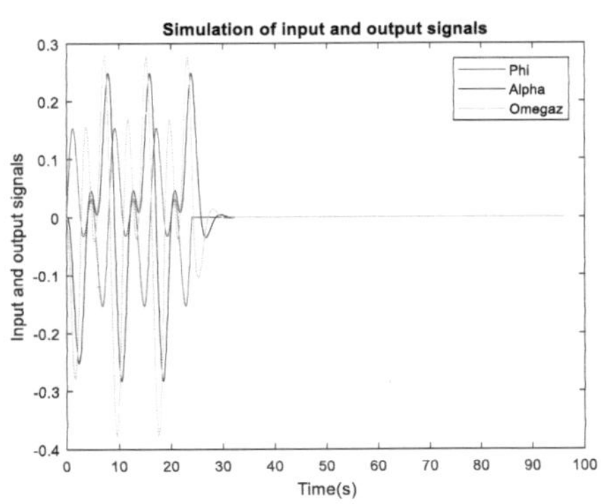

Figure 2. Simulation of the signals without any noises in 24 seconds.

The normally distributed random variables which have zero mean and the standard deviation (0.2) are used as the measurement noises in this work since it is obvious that the measurement noise also influences the identification process [6].

4. Least square method

A typical identification criterion typically has a form of minimization of a function of the prediction errors. The function is usually based on a distance metric. For example, least squares methods minimize the Euclidean distance between the predicted and observed values (squared 2-norm of the prediction errors). Other factors such as quality of coefficient estimates, number of coefficients can be factored into the objective function. In order to carry out the identification of the aerodynamic coefficient, the least square method (LSM) (4.2) is used. This least square method (LSM) is more effective for the linear systems. The LSM is used in the time domain and it is a method for estimating linear parameters in a relation such as;

$$Y = Xa$$

$$\begin{bmatrix} Y(t_1) \\ Y(t_2) \\ \vdots \\ Y(t_N) \end{bmatrix} = \begin{bmatrix} 1 & x_{1t_{(1)}} & x_{2t_{(1)}} & x_{3t_{(1)}} \\ 1 & x_{1t_{(2)}} & x_{2t_{(2)}} & x_{3t_{(2)}} \\ & & \cdots & \cdots \\ 1 & x_{1t_{(N)}} & x_{2t_{(N)}} & x_{3t_{(N)}} \end{bmatrix} \begin{bmatrix} a_1 \\ a_2 \\ \vdots \\ a_N \end{bmatrix} \quad (5)$$

In general form for an arbitrary instant t_i it may be expressed as follows;

$$Y(t_i) = a_0 + a_1x_1(t_i) + a_2x_2(t_i) + a_3x_3\,(t_i),$$
$$i = \;\; 1,2, \ldots N,$$

Where, a_0, a_1, a_2, a_3 — coefficients to be identified; N — number of samples.

Where Y is the measured output, X contains the regressors and a contains the coefficients to be estimated. The parameter estimate can then in principle be calculated as

$$\hat{a} = [\textstyle\sum_{i=1}^{N} X_i^T X_i]^{-1} \sum_{i=1}^{N} X_i^T Y_i = (X^T X)^{-1} X^T\, Y \quad (6)$$

Where, $\hat{a}$ — estimates of the coefficients; X — matrix for object model; Y — matrix for output signal

For the estimation of coefficients, it is necessary to form the object model. Let us consider every equation of the object model separately.

5. Sliding-window approach

Given a discrete time series stream N, the sliding window technique examines the (k) most recent data points and moves (s) steps along the time axis as new measurements arrive, where k and s are referred to as window and step size. This technique has the advantage that it does not need to store the never-ending stream of data, but it also implies that measurements can only be considered for further data analysis as long as they are located within the current window.

In order to analyze the accuracy of aerodynamic coefficients estimation, sliding window approach is applied. In this work the sliding window method is applied for analyzing the accuracy of estimation both in the presence and the absence of the input signal which is formed by the pilot control. For the mth, n^{th} sliding window of the least square estimation, observation vector $Y_{m,n}$ and observation matrix

$X_{m,n}$ are formed. Note that, it is necessary to limit the number of data vectors $Y_{m,n}$ and $X_{m,n}$ from which the least square estimation matrix is calculated. Then the mth, n^{th} sliding window for least square estimation can be expressed as follow;

$$\widehat{a}_{m,n} = \left(X_{m,n}{}^{T} X_{m,n}\right)^{-1} X_{m,n}{}^{T}\, Y_{m,n} \quad (7)$$

Where, $m = 0,\ (k + s),\ (2k + 2s), (3k + 3s), \ldots$, N–k; $n = (k), (2k + s), (3k + 2s),\ (4k + 3s), \ldots$, N; k = sliding window size; s = shifting interval; N = the length of processing time.

The window size is adjustable according to actual needs, based on the length of processing time. This can guarantee the speed of calculation and can reduce the system resources possession too. At this time, the parameter estimation is related to the current data sample, the past time of data has no effect on parameter estimation, and this can ensure the instantaneity and accuracy of the parameter estimation.

The sliding-window model considers the window with specific time interval and the shifting interval.

6. Estimation of aerodynamic coefficients

The input matrices and output vectors for the least square window (7) must be formed to perform parameter estimation. In order to form the input matrix for estimating the force coefficients, the input signals (φ_t), angle of attack (3), and overload are used. For the formation of output vector, the overload is used, which is expressed as follow;

$$n_y(t_i) = \left[\frac{V}{g}\left(Y^{\alpha}\alpha(t_i) + Y^{\varphi}\varphi(t_i)\right)\right] + R(t_i) \quad (8)$$

Where, $n_y(t_i)$ — overload with specific noise for time instant (t_i); V — airspeed (ms^{-1}); g — gravitational acceleration (ms^{-2}); $R(t_i)$ — normally distributed random error.

For the purposes of estimation of three moment coefficients, the input matrix is formed as the first one was formed for two force coefficients, and the angular velocity of pitch (4) is presented in the form;

$$\frac{d\omega(t_i)}{dt} = M_z^{\alpha}\,\alpha(t_i) + M_z^{\omega_z}\omega_z(t_i) - M_z^{\varphi}\varphi(t_i) \quad (9)$$

The output vector Y consists of the estimates for angular velocity of pitch time derivative.

$$\frac{d\omega(t_i)}{dt} = \frac{\omega(t_{i+1}) - \omega(t_{i-1})}{2\Delta t}$$

The simulation data and the processing of this data by least square method (LSM) were repeated many times for each of sliding windows to investigate statistically the accuracy of the parameter estimation, the influence of measurement noises and how influence the presence and absence of input signal on the estimation process. All the signals involved were distorted with simulated noises, normally distributed discrete random variables. For each of aerodynamic parameter, the means and standard deviations of the estimates were determined and then relative errors of these means of aerodynamic parameters were calculated in order to statistically prove the accuracy of aerodynamic parameters estimation by least square method.

It should be explained that a widely known formula (6) associated with the information matrix $(X^T X)^{-1}$ for the LSM estimates errors dispersion, is valid in the case of output measurements noises only. In this work, we assume that all the signals used for identification, are affected by measurements errors. For this reason we apply the statistical simulation in order to investigate the accuracy of the estimated parameters.

7. Analysis of relative errors of the estimated coefficients

After the process of estimation, we generally assume that some exact or true value exists based on how we define what is being estimated. While we may never know this true value exactly, we attempt to find this ideal quantity to the best of our ability with the time and resources available.

In this step, the relative errors (6.1) of the estimated aerodynamic coefficients are determined based on the presence and absence of input signal. It is obvious, that the measurement noises influence the accuracy of the estimated parameters and the accuracy of estimation also depends on the conditions (presence and absence) of input signal. The relative errors of the estimated parameters are shown in Fig. 3.

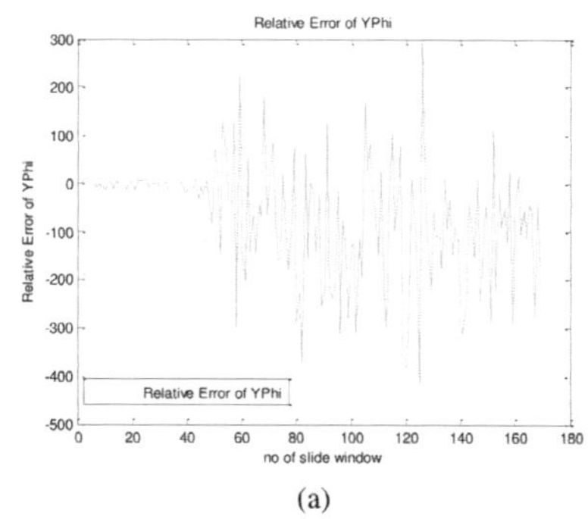

(a)

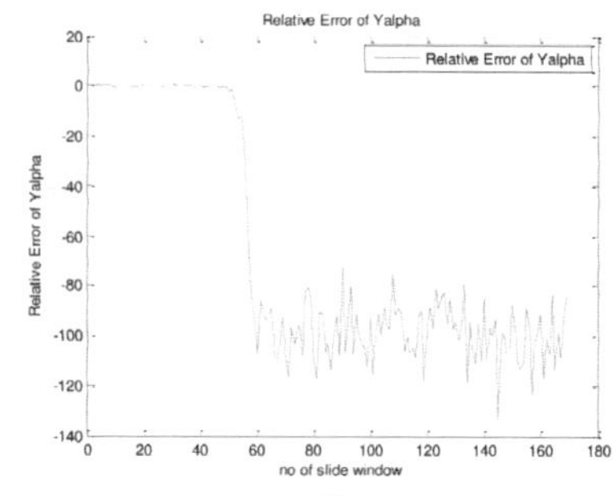

(b)

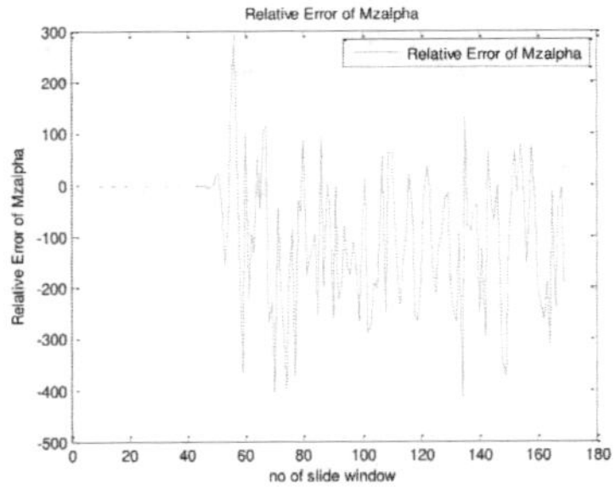

(c)

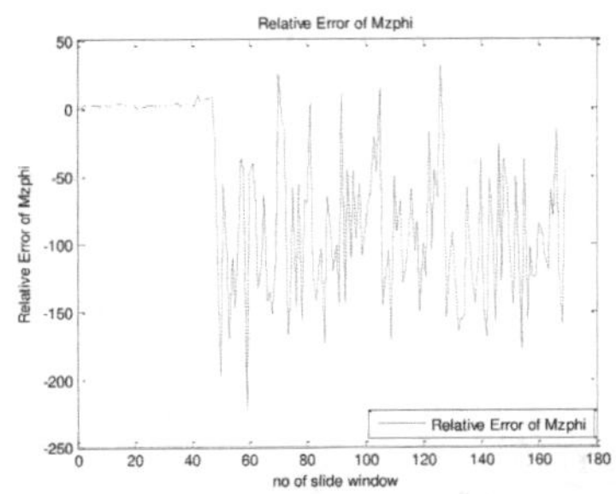

(d)

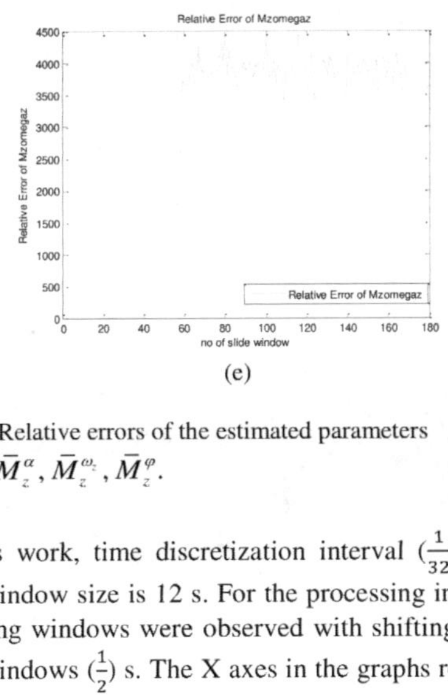

(e)

Figure 3. Relative errors of the estimated parameters $\bar{Y}^{\alpha}, \bar{Y}^{\varphi}, \bar{M}_z^{\alpha}, \bar{M}_z^{\omega_z}, \bar{M}_z^{\varphi}$.

In this work, time discretization interval $(\frac{1}{32})$ s and the sliding window size is 12 s. For the processing interval 84 s 169 sliding windows were observed with shifting interval of sliding windows $(\frac{1}{2})$ s. The X axes in the graphs represent the sliding windows numbers, that is the window starting time, and relative errors of the estimated parameters are expressed on the Y axes of the graphs. Fig. 3 shows that estimation errors keep being small while the windows starting time are within first 22 s. According to fig.2 , the input signal length is 24 s. It means that we may obtain accurate estimates even when the input signal is acting at a part of the sliding window – 2 s out of 12 s that is 17% of the window full length.

8. Conclusion and discussion

According to the results, it could be easily seen that the higher accuracy of the identification is obtained during the presence of the input signal even at a small part of the processing window more than 17 % of its full length.

Acknowledgements

This work was supported by Russian Foundation for Basic Research (RFBR), project 18-08-00921-a.

References

[1] Vasil'chenko K K, Leonov V A, Pashkovskii I M, and Poplavskii B K 1996 *Flight Testing of Aircrafts* (Moscow: Mashinostroenie)
[2] Klein V and Morelli E A 2006 *Aircraft system identification: Theory and Practice* (USA, Reston: AIAA)
[3] Ovcharenko V N 2017 *Aerodynamic characteristics identification of aircrafts by flight data* (Moscow: Publishing house of MAI)
[4] Larsson R 2013 *System Identification of Flight Mechanical Characteristics*. Linköping Studies in Science and Technology, Licentiate Thesis (Sweden: Linköping University)
[5] Korsun O N 2003 *Automation and remote control* **64** 772–81
[6] Moung Htang Om, Kyaw Zin Latt and Karapetyrn T S 2017 Estimation of aerodynamic parameters in condition of measurement *ITM Web of Conferences* **10** 01007
[7] Moung Htang Om and Kyaw Zin Latt 2017 *Cloud of Science* **4** 636–49
[8] Moung Htang Om 2018 *Aerodynamic Parameters Identification in Aircraft Longitudinal Motion* (LAP Lambert Academic Publishing)

Modeling Systems Based on the Assessment of Weak Symmetry Breaking in Reconstructed Attractors

Evgeny Nikulchev[1]

[1] MIREA – Russian Technological University, Moscow, Russian Federation

E-mail: nikulchev@mail.ru

Received xxxxxx
Accepted for publication xxxxxx
Published xxxxxx

Abstract

The paper proposes to use the research method for chaotic processes based on the identification of weak symmetry breaking of the restored attractor. It is shown how the calculations results can be used to identify systems. An algorithm for creating finite-difference models has been developed, including: calculating by numerical methods the necessary conditions for the existence of chaos, reconstructing an attractor in a time series, searching for symmetric attractor fragments under conditions of weak symmetry breaking, determining the form of nonlinearities, and parametric identification. The result of the algorithm is a system of finite-difference equations in the state space Criteria for assessing symmetry breaking based on estimates of the divergence of fragments of phase trajectories are introduced. The results of modeling systems with chaotic dynamics are presented.

Keywords: attractor, chaotic processes, symmetry

1. Introduction

Currently, chaotic signals are used in various fields of science and technology — in metallurgy [1], mechanical engineering [2], tasks of automation of technological processes [3, 4], for modeling networks [5, 6] and many others. his indicates the great interdisciplinary significance of scientific results in the field of the theory of chaotic dynamics. In many practical applications, it is convenient for researchers to act on the assumption of the chaotic nature of the observed process. The attractor is easy to recover; there are many numerical methods for estimating dimensions. Methods for the approximate solution of the inverse problem of dynamics are developed, i.e. restoration of a given type of differential equations from experimental data.

However, the assumption of chaotic dynamics is often based only on estimates of the necessary conditions for the existence of chaos, such as an estimate of the senior Lyapunov exponent. The paper presents an algorithm for detecting weak symmetry breaking for the analysis of the chaotic nature of the reconstructed phase space [7, 8]. It is shown how the results of calculations can be used to identify systems.

For the study, a time series $\{y_0, y_1, ..., y_k, y_{k+1}, ...\}$ is used, while it is assumed that the time series is generated by the system, which discretized form has the following form:

$$\begin{aligned} \mathbf{x} &= f(\mathbf{x}_k, \mathbf{x}_0), \\ y_k &= g(\mathbf{x}_k), \end{aligned} \qquad (1)$$

where $\mathbf{x}$ — n-dimensional point in the state space; y_k — observed one-dimensional process; k — discrete time (number); f, g — vector function.

Using discrete models allows operating with data obtained as a result of the experiment, and accumulate no errors associated with the transition to continuous models, however, this complicates the qualitative analysis of processes.

As is known, a phase trajectory is a trajectory depicting how the state of the dynamical system x changes with time t. For discrete systems, states are connected by lines, in accordance with the sequence of samples $k = 1, 2, ...$ The area to which all possible trajectories of the systems evolve (converge) is called an attractor. For discrete systems, as well as for continuous, the attractor is a point, for oscillatory systems — closed trajectories (cycles). For chaotic systems, there is an attractor called strange, in which case the trajectories are contracted, but not to a point, a curve, a torus, but to a certain subset of the phase space.

Currently, there are different points of view on the number and nature of scenarios of transition to chaos:

- Feigenbaum scenario (through a cascade of period doubling bifurcations);
- Afraimovich-Shilnikov scenario (Ruel-Tackens model) (through the destruction of a two-dimensional torus or through the destruction of a closed invariant curve);
- Pomo-Manneville scenario (alternating in time of almost regular oscillations with intervals of chaotic behavior).

The study of scenarios for transition to chaos plays an important role in practice, since it makes it possible in some cases to predict the possibility of a chaotic behavior of a dynamic system by the control parameters change. When developing the method, the second type of scenario was used, which allows the use of methods of qualitative nonlinear dynamics.

The attractor is an invariant characteristic of the system, i.e. it is stored during transform actions. Dynamic systems for which the n-dimensional phase volume decreases are called dissipative; if the phase volume is preserved, then such systems are called conservative. Conservative systems always have at least one conservation law. Conservation laws are determined by symmetries allowed by the system.

The problem considered in the paper consists in constructing a model of the form (1) from the observed time series.

2. Modeling systems with chaotic dynamics based on the assessment of weak symmetry breaking in a reconstructed attractor

A method for modeling systems with chaotic dynamics was developed. This method allows finding a solution to the inverse problem of dynamics on a minimal invariant manifold in the class of affine systems.

The essence of the method is as follows. The initial data is the time series generated (presumably) by a system with chaotic dynamics.

- The necessary conditions for the existence of chaos are calculated by numerical methods — the largest Lyapunov exponent (for chaotic dynamics it must be greater than zero), etc.
- The attractor is restored by the Packadra method (delay method).
- If the assumption of chaotic dynamics is confirmed, then research continues.
- Based on the analysis of the reconstructed attractor, the presence of symmetries is checked (in conditions of weak symmetry breaking) [9].
- Using the accepted symmetries, the form of equations in the minimal invariant manifold is constructed using the Hausdorff-Lee formula [10].
- The structure of equations is parametrically identified.
- The qualitative discrepancy between the dynamics of the model and the initial series is estimated.

The output is a model in the form of finite difference equations.

An algorithm for searching for symmetric sections of reconstructed trajectories under conditions of weak symmetry breaking was developed [10]. It should be noted that in this formulation, the problem of finding almost symmetric fragments is NP-complete.

3. Chaotic signal modeling and generation

It was shown in [11] that robust chaos can occur in piecewise-smooth systems. This result facilitates the task of modeling systems with chaotic dynamics. Attempts to carry out parametric identification of a model with arbitrarily selected types of nonlinearities, as a rule, do not find a solution at all. If a solution is found (which is rather an accident than a natural phenomenon), then the solutions will be naturally sensitive to minor changes in parameters. Indeed, in many physical phenomena, chaos occurs with a slight change in the parameter, however, in technical applications, the use of a chaotic regime usually requires a search for less sensitive ones, i.e. robust models. This requires the very essence of experiments with technical systems, as well as errors in calculations and measurements. In addition, for the problems of control synthesis, it is efficient to use the chaotic properties of systems, but provided that the chaotic regime does not change the qualitative behavior in a certain range of parameters.

It is proposed to use the model in the following form:

$$\begin{aligned} \mathbf{x}_{k+1} &= \mathbf{A}\mathbf{x}_k + \mathbf{B}\Psi u_k, \\ \mathbf{y}_k &= \mathbf{C}\mathbf{x}_k, \end{aligned} \tag{2}$$

Here, the nonlinear function Ψ is determined from the found symmetric properties of the phase trajectory according to the Hausdorff-Lee formula; u_k is a discretized piecewise continuous function of time (for example, piecewise linear).

This type of model representation allows the use of parametric identification methods, for example, the least squares method implemented in the System Identification

Toolbox in MATLAB. As parameters of the ident function, the investigated $\{y_k\}$ (output) and the generated piecewise continuous series $\{\Psi u_k\}$ (input) are used. The output of the function is the matrix **A**, **B**, **C** of the found dimensions. Note that the dimension of the state space obtained as a result of identification of the system is always equal to the numerical estimates of the initially reconstructed attractor.

Here, chaos is generated as a periodic change in the linear trajectory, preventing the dynamics from arriving at a stable or periodic trajectory. Of course, not in all applications such models seem convenient, but for control systems the use of the form (2) allows the use of well-known methods of synthesis and parameter estimates. In addition, for engineering tasks, ensuring the security of information systems where chaotic signals are widely used, this method allows generating robust chaos with desired properties (type of attractor, degree of symmetry breaking, etc.).

As an example, we present the type of models obtained for the cooling aluminum ingots in the aviation industry:

$$A=\begin{pmatrix} 0{,}998 & 0{,}006 & -0{,}009 & 0{,}007 & -0{,}001 & -0{,}007 \\ 0{,}008 & 0{,}913 & 0{,}238 & 0{,}142 & 0{,}091 & -0{,}215 \\ 0{,}006 & 0{,}132 & 0{,}091 & -0{,}928 & -0{,}459 & -0{,}038 \\ -8{,}12\cdot10^{-5} & 0{,}008 & -0{,}177 & 0{,}356 & -0{,}956 & 0{,}097 \\ -0{,}001 & -0{,}06 & 0{,}803 & 0{,}203 & -0{,}145 & 0{,}139 \\ 0{,}002 & -0{,}006 & 0{,}006 & -0{,}003 & 0{,}010 & 0{,}661 \end{pmatrix};$$

$$\Psi=\exp(10k)\left(-7{,}24\cdot10^{-4};-0{,}087;0{,}209;-0{,}074;-0{,}440;-0{,}061\right)^T$$

$$C=\begin{pmatrix} 129{,}8 & -144{,}2 & 18{,}43 & -23{,}21 & -3{,}233 & -4{,}911 \\ 582{,}1 & 4{,}995 & 2{,}230 & -1{,}596 & 2{,}587 & 4{,}100 \end{pmatrix}.$$

The attractor is shown in Fig. 1. A comparison of the simulated output process with real data is shown in Fig. 2.

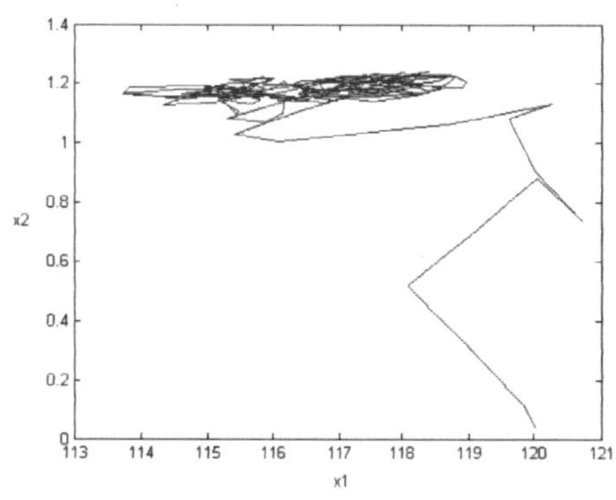

Figure 1. Reconstructed Attractor.

It can be seen that the model exhibits complex dynamic behavior that is adequate to the heat transfer process under consideration.

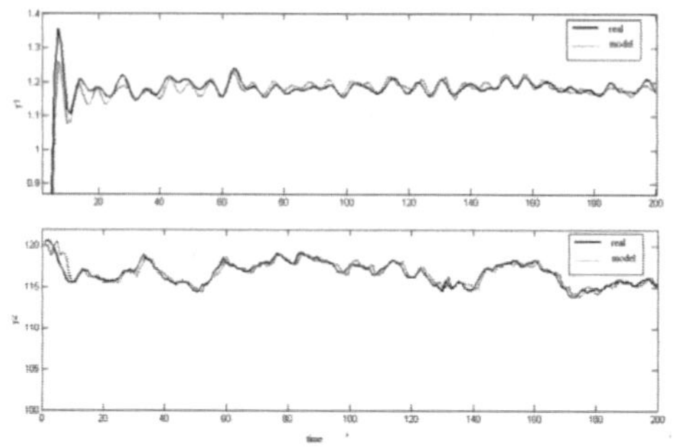

Figure 2. Comparison of real data with the model.

4. Conclusions

A method for creating mathematical models of chaotic processes is developed. The obtained models can be used to build robust chaos, as well as to generate chaos with specified qualitative properties.

The use of dynamic models of chaotic processes makes it possible to provide predictive proactive management for the dynamic allocation of resources in complex infrastructures. An important task is the choice of the shape of the model. Using the proposed form will allow us to obtain computationally reliable models that, on the one hand, reflect the necessary qualitative properties of nonlinear systems, and on the other, are not sensitive to insignificant changes in parameters.

References

[1] Volkova V E 2015 *IOP Conference Series: Materials Science and Engineering* **96** 012072
[2] Hosseinloo A H, Slotine J J and Turitsyn K 2018 *Journal of Vibration and Control* **12** 2532–41
[3] Kiseleva M A, Kuznetsov N V and Leonov G A 2016 *IFAC-PapersOnLine* **14** 51–5
[4] Tebueva F B, Kopytov V V, …, and Demirtchev N G 2018 *Journal of Theoretical & Applied Information Technology* **8** 2197
[5] Karpukhin A, Griciv D and Nikulchev E 2018 *Journal of Theoretical & Applied Information Technology* **5** 5080–93
[6] Sigov A S, Andrianova E G, Zhukov D O, Zykov S V and Tarasov I E 2019 *Russian Technological Journal* **1** 5–37 DOI: 10.32362/2500-316X-2019-7-1-5-37
[7] Grigoriev R O 2000 Physica D: Nonlinear Phenomena **140** 171–92 doi: 10.1016/S0167-2789(00)00014-2
[8] Jiang J J, Zhang Y and Stern J 2001 *The Journal of the Acoustical Society of America* **4** 2120–8 doi: 10.1121/1.1395596
[9] Nikulchev E V (2015). Generation of robust chaos in the invariant centre manifold *2015 International Conference "Stability and Control Processes" in Memory of VI Zubov (SCP)* pp 290–1

[10] Nikulchev E 2014 Robust chaos generation on the basis of symmetry violations in attractors *2014 2nd International Conference on Emission Electronics (ICEE)* pp 59–61

[11] Banerjee S, Yorke J A and Grebogi C 1998 *Physical Review Letters* **80 14** 3049

IOP Conf. Series: Materials Science and Engineering **714** (2020) 012023 doi:10.1088/1757-899X/714/1/012023

Algorithm Development for Estimating Projections of Wind Velocity Using Measurements of Airspeed, Angle of Attack and Sideslip

Moung Htang Om[1], Kyaw Zin Latt[1] and A I Zhelannikov[2]

[1] Moscow Aviation Institute, Moscow, Russia
[2] Central Aerohydrodynamic Institute, Zhukovsky, Russia

E-mail: moughtangom50@gmail.com

Received xxxxxx
Accepted for publication xxxxxx
Published xxxxxx

Abstract

The article considers the problem of estimating wind velocity in flight. The method proposed in this paper provides estimates for three projections of wind velocity in a normal Earth coordinate system using satellite navigation data (SNS), as well as on-board barometric measurements of flight speed. It is assumed that the wind has a constant velocity and direction for the flight interval, the duration of which is 30–31 s. The proposed algorithm can provide measurement of three projections of wind velocity for interval 0.5 s, which compares favourably with other options when the required observation interval was tens of seconds.

Keywords: estimation, maneuver, wind velocity, angle of attack, sideslip, airspeed

1. Introduction

In aeronautical technology, the measurement of atmospheric parameters and aircraft motion parameters is very important during flight tests [1]. To estimate the measurement errors, the system identification method can be utilized [2–5]. Recently, parameter identification algorithms have been proposed for systematic errors in measuring the angle of attack and sideslip angle of an aircraft [6], orientation angles [7] and airspeed [8]. At the same time, determining the actual values of wind velocity during flight test modes is relevant. The possibilities for developing methods of estimating wind speed by using satellite navigation systems in flight tests have expanded significantly, especially in comparison with traditional technologies for optical measurements of the flight path [9]. Accuracy of estimation of wind velocity is necessary for numerous flight test tasks of the aircraft [1–11]. The proposed method in this article provides estimates for 3 projections of wind velocity in the Earth's normal coordinate system using data from satellite navigation systems (SNS), as well as barometric airborne measurements. In this paper, it's proposed an algorithm which uses sliding window model, as well as its accuracy characteristics, taken from the processing of simulation data. The influence of airspeed error on the estimation process of wind velocity is also discussed.

2. Problem formulation

It is assumed that the wind has a constant velocity and direction for a short flight interval which duration is 30 s and 31 s. This means, that for this time interval the values of the wind velocity projections on the axis of the normal earth coordinate system are constant. In order to define the sequence of operations to generate estimates of components of wind speed, it's necessary to create the object model. Equations for projection of airspeed of the aircraft on the normal earth coordinate system are defined as follow:

$$V_{xg\_a}(t_i) = V_{xg\_CHC}(t_i) - W_{xg},$$
$$V_{yg\_a}(t_i) = V_{yg\_CHC}(t_i) - W_{yg}, \quad (1)$$
$$V_{zg\_a}(t_i) = V_{zg\_CHC}(t_i) - W_{zg},$$

where, $V_{xg\_CHC}(t_i), V_{yg\_CHC}(t_i), V_{zg\_CHC}(t_i)$ — the projections values of the airspeed of the aircraft measured by SNS on the axis of the normal earth system; W_{xg}, W_{yg}, W_{zg} — unknown projections values of wind velocity on the axis of the normal earth system to be identified.

Obviously, the modulus of the airspeed vector is given as follow:

$$V_a(t_i) = \sqrt{V^2_{xg\_a}(t_i) + V^2_{yg\_a}(t_i) + V^2_{zg\_a}(t_i)}. \quad (2)$$

To go to the body axes, the corresponding matrix of guide cosines is utilized:

$$\begin{bmatrix} V_{x\_a} \\ V_{y\_a} \\ V_{z\_a} \end{bmatrix} = \begin{bmatrix} \cos\psi\cos\vartheta & \sin\vartheta & -\sin\psi\cos\vartheta \\ \sin\psi\sin y - \cos\psi\sin\vartheta\cos y & \cos\vartheta\cos y & \cos\psi\sin y + \sin\psi\sin\vartheta\cos y \\ \sin\psi\cos y + \cos\psi\sin\vartheta\sin y & -\cos\vartheta\sin y & \cos\psi\cos y - \sin\psi\sin\vartheta\sin y \end{bmatrix} \times \begin{bmatrix} V_{xg\_a} \\ V_{yg\_a} \\ V_{zg\_a} \end{bmatrix}. \quad (3)$$

During the calculations, the orientation angles should be taken according to the output signals of the vertical direction included in the navigation system. A constant yaw error, which is permissible to be considered a constant over a section of 15 ... 80 s, should be included in the vector of identifiable parameters.

The multiplicative error of the measuring channels is taken into account by the steepness coefficients of the aerodynamic angle sensors K_α and K_β.

In the presence of nonlinearity, an additional approximation is introduced, which is not of fundamental difficulty, but it increases the dimension of the problem.

By using the projection of airspeed on the body axes, the formulas for the values of the angle of attack and sideslip can be written as follow:

$$\alpha_u(t_i) = -arctg\left(\frac{V_{y\_a}(t_i)}{V_{x\_a}(t_i)}\right),$$
$$\beta_u(t_i) = -\arcsin\left(\frac{V_{z\_a}(t_i)}{V_a(t_i)}\right), \quad (4)$$

where, $\alpha_u(t_i), \beta_u(t_i)$ — output signals of the measuring channels of attack and sideslp.

The object model is determined by the equations (1)–(4).

The observation model takes the form:

$$z_2(t_i) = V_a(t_i) + C_V + \xi_V(t_i),$$
$$z_2(t_i) = K_\alpha \alpha_u(t_i) + C_\alpha + \xi_\alpha(t_i), \quad (5)$$
$$z_3(t_i) = K_\beta \beta_u(t_i) + C_\beta + \xi_\beta(t_i),$$

where, C_V, C_α, C_β — additive errors of aerometric measuring channels; K_α, K_β — multiplicative error coefficients; $\xi_V(t_i), \xi_\alpha(t_i), \xi_\beta(t_i)$ — random errors of aerometric measurements.

Thus, the identification is aimed at obtaining estimates of the following quantities, where the first three are the coordinates of the wind velocity vector, and the rest specify the systematic errors of the measuring channels:

$$\alpha^T = \left[W_{xg}\ W_{yg}\ W_{zg}\ C_V C_\psi\ C_\alpha\ C_\beta\ K_\alpha\ K_\beta \right].$$

3. Identification Algorithm

The problem of wind velocity estimation may be solved using the maximum likelihood estimation (MLE) algorithm [2, 4, 6]. In general vector form object model and observation model are presented as follows:

$$y(t_i) = f(y(t_i), a, u(t_i)), \quad (6)$$
$$z(t_i) = h(y(t_i), a, u(t_i)) + \eta(t_i). \quad (7)$$

where $y(t)$, $u(t)$ — vectors of object output and input signals which dimensions are n and m respectively; $z(t_i)$ — vector of observation with dimension r; a — vector of unknown parameters which have to be estimated; $\eta(t_i)$ — noise of observations, which is a vector of normal discrete random sequence with zero mean and known variance matrix $R(t_i)$.

It is assumed that input signal $u(t)$ is a known function of time. The initial conditions $y(t_0)$ are also assumed to be known.

Noise observations are normal and independent random vector variable. It is known that the maximum likelihood method under the specified assumptions about the properties of noise leads to unbiased and efficient estimates [2]. The minimized functional of the maximum likelihood method is expressed in the following form:

$$J(a) = \sum_{i=1}^{N}((z(t_i) - h(y(t_i), a, u(t_i)))^T \times$$
$$\times R^{-1}(t_i)((z(t_i) - h(y(t_i), a, u(t_i))), \quad (8)$$

It is easy to see that (8) is a functional of the least squares method with the matrix of weight coefficients $R(t_i)^{-1}$.

For the minimization of (8), it is proposed to use the modification of the classical Newton method

$$a_{k+1} = a_k - \left(\frac{d^2 J(a_k)}{da_k^2}\right)^{-1} \frac{dJ(a_k)}{da_k}$$

where;

$$\frac{dJ(a_k)}{da_k} = -2\sum_{i=1}^{N} \frac{dz^T(t_i, a_k)}{da_k} \cdot R^{-1}(t_i)(z(t_i) - z(t_i, a_k)), \quad (9)$$

$$\frac{d^2J(a_k)}{da_k^2} \approx 2\sum_{i=1}^{N} \frac{dz^T(t_i,a_k)}{da_k} \cdot R^{-1}(t_i)\frac{dz(t_i,a_k)}{da_k}. \qquad (10)$$

The derivative estimates are determined numerically for the discrete time $t_i, i=\overline{1,N}$ according to the formulas:

$$\frac{dz(t_i,a)}{da} = \left[\frac{\partial z(t_i,a)}{\partial a_1} \quad \frac{\partial z(t_i,a)}{\partial a_2} \quad \cdots \quad \frac{\partial z(t_i,a)}{\partial a_p} \right]_{(r\times p)}, \qquad (11)$$

$$\frac{\partial z(t_i,a)}{\partial a_j} = \frac{z(t_i,a+\varepsilon e_j)-z(t_i,a)}{\varepsilon}. \qquad (12)$$

where e_j — vector of dimension p, all of elements of which are equal to zero except j element and j element is equal to 1; ε — a small number, usually specified at the level of 0.001 ... 0.1% from the nominal parameter values.

Estimates $z(t_i,a), i=\overline{1,N}$ are determined by numerical solution of equations of the object and observations in $\eta(t_i)=0$. Identification is finished when we meet the condition $|a_{k+1}-a_k| < \delta|a_k|$, where $\delta = 0,005$.

4. Estimating projections of wind velocity using sliding window

The capabilities of this algorithm were investigated according to bench simulation. According to the research [kzl] the dimension of the observation vector, that is, an increase in the amount of information used to identify wind speed, will increase the accuracy of the estimates. In this case, random errors were modeled. The main attention was paid to assessing the influence of the type of maneuver and the duration of the moving interval. Such maneuver as pitching dachas, "barrel" is considered. In this version of the algorithm, it was possible to significantly reduce the duration of the sliding interval. In this work, investigation was performed in sliding interval of 0.5 and 1.0 s. To determine the influence of the motion parameters, the beginning of the sliding interval was shifted sequentially throughout the processing area with a step of 1 s.

5. Maneuver "barrel"

The results for the "barrel" maneuver are presented in tables 1 and 2, as well as in fig. 1 and 2.

Table 1. Relative errors of three projections of wind velocity estimation for the "barrel" maneuver with sliding interval 0.5 s in processing time 31 s.

Processing interval starting time, s	Relative estimation errors for W_x,%	Relative estimation errors for W_z, %	Relative estimation errors for W_y, %
0	–2.742857143	–1.1	23.5
1	–3.742857143	–2.06	0.4
2	–3.2	–0.44	–23.4
3	–2.714285714	0.22	0.75
4	–2.857142857	–0.4	17.65
5	–2.642857143	–0.7	7.45
6	–2.228571429	–0.26	–1.3
7	–2.8	–0.12	–12.85
8	–2.671428571	–0.16	–5.95
9	–2.814285714	–0.46	11.75
10	–3.085714286	–0.5	6.6
11	–2.685714286	–0.4	–0.1
12	–3.171428571	–1.54	21.35
13	–3.957142857	–7.16	18.8
14	–5.028571429	–0.06	13.7
15	–3.814285714	1.16	3.55
16	–3.171428571	–0.18	2.75
17	–3.085714286	3.74	4.05
18	–3.371428571	13.86	–4.15
19	2.657142857	1.82	–12.15
20	–3.557142857	1.42	3.55
21	–0.285714286	–1.94	–7.45
22	–2.2	–3.28	0.95
23	–3.814285714	–2.3	–15
24	–2.685714286	–3.54	–27.5
25	–2.457142857	–2.98	–2.5
26	–3.014285714	–3.08	–4
27	–2.928571429	–2.84	–5.65
28	–2.628571429	–2.38	–3.2
29	–1.657142857	–1.84	5.15
30	–2.214285714	–2.3	12.8
31	–2.071428571	–2.2	17.65

Table 2. Relative errors of three projections of wind velocity estimation for the "barrel" maneuver with sliding interval 1.0 s in processing time 30 s.

Processing interval starting time, s	Relative estimation errors for W_x,%	Relative estimation errors for W_z, %	Relative estimation errors for W_y, %
0	–2.128571429	–0.74	30.2
1	–2.814285714	–1.24	–12.75
2	–2.5	0.44	–20.6
3	–1.957142857	0.68	4.3
4	–2.342857143	0.02	15.15
5	–2.457142857	0.04	5.1
6	–2.357142857	0.3	–5.3
7	–2.2	0.56	–13.7

8	−1.914285714	0.32	0.15
9	−2.042857143	0.02	8.95
10	−2.028571429	0.14	3.45
11	−2.357142857	0.18	4.3
12	−2.714285714	−3.14	27.75
13	−3.614285714	−6.66	13.8
14	−3.814285714	−0.88	9.6
15	−3.7	−0.42	4.65
16	−3.285714286	0.52	5.55
17	−4.585714286	7.14	8.1
18	−0.142857143	9.4	−4.95
19	1.485714286	1.82	−6.85
20	−1.328571429	2.2	−5.05
21	0.042857143	−1.88	−9.25
22	−2.6	−2.02	3.6
23	−3.185714286	−1.8	−27.05
24	−2.3	−2.4	−20.6
25	−2.585714286	−2.34	−0.7
26	−2.571428571	−1.88	−4.9
27	−2.514285714	−1.7	−6.15
28	−2.128571429	−1.46	−0.5
29	−1.757142857	−1.34	8
30	−2.042857143	−1.54	13.45

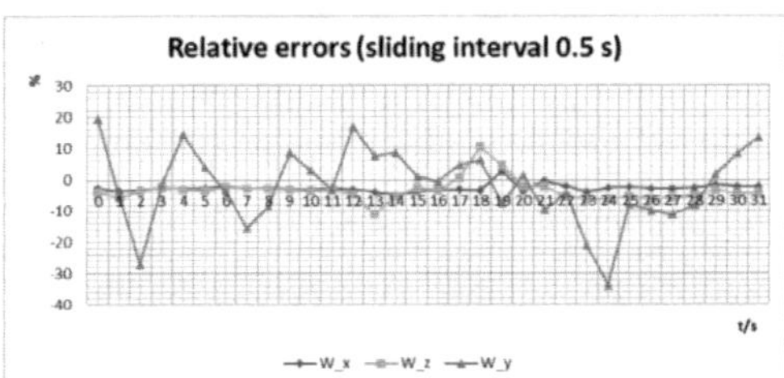

Figure 1. The relative estimation error on the start time of the processing, the sliding interval is 0.5 s, the duration of the processing time is 31 s.

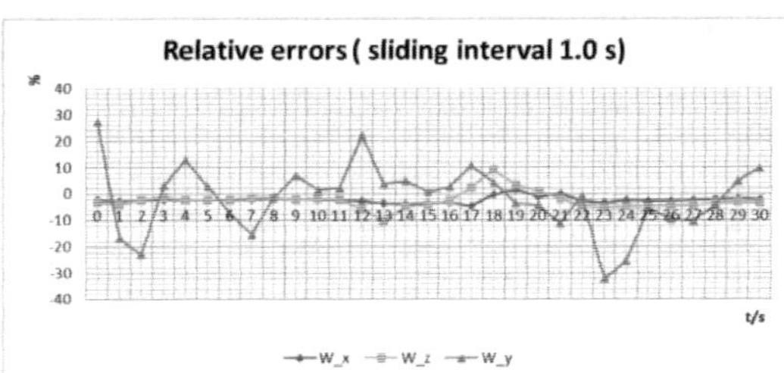

Figure 2. The relative estimation error on the start time of the processing, the sliding interval is 1.0 s, the duration of the processing time is 30 s.

It's obvious, the errors in estimating horizontal projections generally do not exceed 5% over the entire area of the maneuver, and the errors in estimating the vertical component are ± 10%.

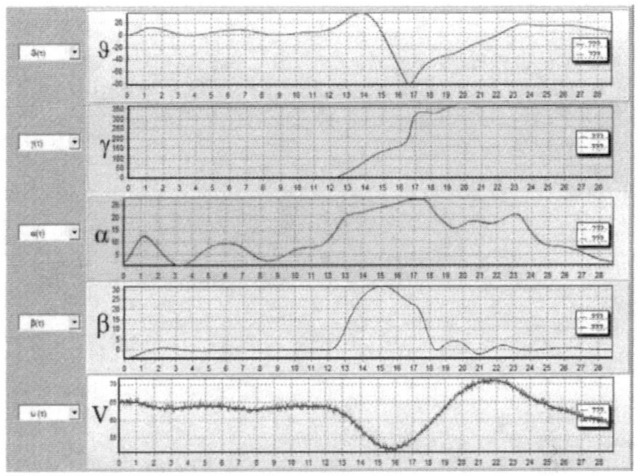

Figure 3. Values of the main flight parameters for the "barrel" maneuver, total maneuver time 31 s.

A comparison of the graphs with Fig.3, which shows the change in the signals during the maneuver, shows that a certain increase in errors occurs at the time of energetic maneuvering, at high speeds of changes in flight parameters.

6. Conclusion

The comparison of the graphs for the sliding window duration of 0.5 s and 1.0 s shows insignificant differences, of the order of 2–3%. This means that the algorithm can provide measurement of three projections of wind velocity for 0.5 s, which compares favorably with other options when the required observation interval was tens of seconds. The short duration of the sliding interval also allows to quickly monitor the change in wind during the flight.

Acknowledgements

This work was supported by Russian Foundation for Basic Research (RFBR), project 18-08-00921-a.

References

[1] Vasil'chenko K K, Leonov V A, Pashkovskii I M and Poplavskii B K 1996 *Letnye ispytanija samoletov* (Moscow: Mashinostroenie)

[2] Klein V and Morelli E A 2006 *Aircraft system identification: Theory and Practice* (USA, Reston: AIAA)

[3] Korsun O N, Poplavsky B K 2014 Approaches for flight tests aircraft parameter identification *29th Congress of the International Council of the Aeronautical Sciences, ICAS 2014* 2014–0210

[4] Jategaonkar R V 2006 *Flight vehicle system identification: A time domain methodology* (USA, Reston: AIAA)
[5] Luchtenburg D M, Rowley C M, Lohry M W, Martinelli L and Stengel R F 2015 *Journal of Aircraft* **52** 890–5
[6] Ovcharenko V N 2017 *Identifikacija ajerodinamicheskih harakteristik vozdushnyh sudov po poletnym dannym* (Moscow: Izdatel'stvo MAI)
[7] Korsun O N and Poplavskii B K 2011 *Journal of Computer and Systems Sciences International* **50** 130–43
[8] Bulgakov V V, Korsun O N, Kulabukhov V S, Stulovskii A V and Timofeev D S 2016 *Journal of computer and systems sciences international* **55** 150–61
[9] Korsun O N, Nikolaev S V and Pushkov S G 2016 *Journal of Computer and Systems Sciences International* **55** 446–57
[10] Gumarov S G and Korsun O N 2011 *Measurement Techniques* **3** 281–6
[11] Moung Htang Om and Kyaw Zin Latt 2017 *Cloud of Science* **4** 636–49

Speech Recognition Using Convolutional Neural Networks on Small Training Sets

A V Poliyev[1] and O N Korsun[1]

[1] State Research Institute of Aviation Systems, Moscow, Russia

E-mail: marmotto@rambler.ru

Received xxxxxx
Accepted for publication xxxxxx
Published xxxxxx

Abstract

The algorithm of separate words automatic recognition based on convolutional neural networks is developed and presented in this paper. Distinctive feature of this algorithm is the training on sets consisting of only hundreds or thousands of samples. Therefore, important problem is the selection of optimal architecture for neural network, which was firstly proposed and tested. After that, four different cases for recognition were researched: speaker-dependent recognition without noise, speaker-independent recognition without noise, speaker-dependent recognition with noise, speaker-independent recognition without noise. Finally, we analyse the experiment results that showed good results for all cases of interest.

Keywords: man-machine interface, speech recognition, convolutional neural networks

1. Introduction

A reliable and rational man-machine interface is an important task of modern engineering [1–3]. In recent years in avionics we witness a considerable growth of attention to audio interfaces and speech recognition [4–9], since their successful implementation may improve the flight safety and reduce the pilots' workload [1, 9]. For example, on Eurofighter Typhoon since 2005 a speaker-dependent system based on the comparing with the patterns is employed.

The principal requirement for speech interface in avionics is high probability of correct word recognition even under strong noises, because an error in aircraft control may be critical for the flight safety. In addition, the time complexity of the algorithm is an important point for onboard software.

In this paper one proposes the algorithm based on convolutional neural networks on small training sets which shows a good performance for recognition of separate words. The description of the method along with proposed structure of used convolutional neural network is presented in theoretical part. The description of employed dataset and th results of the method for several cases are presented in the experimental part.

2. Speech signal parametric portrait in automatic speech recognition

The traditional methods for automatic recognition of speech commands are based on the time-spectral quantisation of an input word record. The transform used in this article obtains a time-spectral parametric portrait and includes the following steps. In the beginning, a speech signal is divided into equal intervals of 20–50 ms. After that, one uses the standard digital signal processing procedures, such as amplification of high-frequency signal components, interval weighing by Hannah window, fast Fourier transform, averaging over frequencies and logarithm of spectral densities [10]. Each of these intervals is projected into 30–40 frequency bands [1]. As a result, for each of the time intervals bands we obtain the estimates of spectral densities logarithms in 30–40 frequency bands [4, 7, 9].

These parameters, presented in a form of a matrix, we call the parametric portrait of the word. The columns of this portrait characterize the spectral decomposition of the speech signal for each time interval.

3. Development of the algorithm of convolution neural network

This section describes the development of algorithms for automatic recognition of speech commands based on convolutional neural networks. When conducting experiments on neural networks, parametric portraits of small sizes were used, and the best results for word recognition were obtained on portraits with 18 frequency bands and 25 time intervals.

To train neural networks of forward propagation, it is common to use the back propagation method. There are batch, stochastic and mixed, i.e. mini-batch implementations of this method. The latest implementation is the most used at the moment and will be applied in this work. Its features include some randomness in the learning process, which leads to slightly different recognition results with the same training parameters. Moreover, in some neural networks regularization is used, which also introduces a small randomness in the learning process. From this we can conclude that all the results are approximate and may not be reproduced exactly at repeated launches.

The simulation of artificial neural networks was carried out in a software package written in the Python programming language using the TensorFlow and TFLearn libraries. TensorFlow is an open source low-level library designed to solve the problems of construction and training neural networks [11]. TFLearn is a high-level library for simplifying work with neural networks; it is a wrapper for the TensorFlow library [12]. Both libraries can perform calculations both on the central processor and on the graphical processor. The use of the latter is in priority, since such processors contain a lot of low-performance cores, which are well suited for the process of training of neural networks, the algorithm of which is efficiently parallelized. To effectively use the graphics processor in the program, the software-hardware architecture of parallel computing CUDA is used, actively using mentioned above plug-in libraries.

In the beginning, there was provided a search for neural network architectures most suitable for use in speech recognition problem. Traditional single and double layer perceptrons showed unsatisfactory recognition results, with the number of errors approximately equal to 30–40%. Further, new architectures of artificial neural networks, mainly recurrent networks and deep learning networks, were tested.

The input array is a two-dimensional parametric portrait. Here, unlike single-layer and multi-layer perceptron, there is no need to convert a two-dimensional parametric portrait into a one-dimensional array. Next comes the convolutional layer with 32 convolutional filters of size 3×3. Then follows a subsample layer with a kernel size of 2, that is, each section of each feature map of size 2×2 is converted to one element in a new feature map. After that, there is a convolutional layer with 64 filters, each of which is applied to all feature cards and at the output of this layer 64 feature cards are obtained. Next is another subsample layer with a kernel size of 2, followed by three fully connected layers with 128, 256 and 20 elements, respectively, where 20 is the number of recognizable words. Thus, the total number of parameters of this neural network is $(3 \times 3 \times 1 + 1) \times 32 + (3 \times 3 \times 32 + 1) \times 64 + (64 \times 3 \times 5 + 1) \times 128 + (128 + 1) \times 256 + (256 + 1) \times 20 = 179'988$.

In convolutional layers, as well as in multi-layer perceptron, the ReLU activation function is used. The first two fully connected layers use the hyperbolic tangent as an activation function, which is given by the equation $f(x) = \tanh(x) = \frac{e^x - e^{-x}}{e^x + e^{-x}}$. Also, after both hidden, fully connected layers, regularization is applied to prevent overfitting of the network. For the last fully connected layer, the softmax function is used to normalize the obtained output probabilities. To train this type of neural network, cross entropy is used as a loss function. The learning process uses the Adam optimization for stochastic gradient descent.

The number and sizes of filters shown above, as well as the dimensions of the final fully connected layers are taken from empirical recommendations for image recognition problem solved by convolutional neural networks. A parametric portrait is a some kind of image, so it was decided to use similar parameters.

An important issue is the use of convolutional neural networks for small training sets, the size of which in our case is from 300 to 4900 samples. It is also worth noting that there is only 20 classes for word recognition problem. The number of elements of each class in the training set is not less than 30.

Although initially convolutional neural networks were trained on large sets, recently encouraging results have been obtained on small training sets. For example, [13] used 27000 elements from 121 unbalanced classes for image recognition, while [14] used the CIFAR-10 database with 50000 elements from 10 classes in the training set [15]. At the same time, under certain conditions, convolutional neural networks show good results when training on small sets containing from 100 to 1000 elements, even with the number of trained parameters about several hundred thousand [16]. The number of gradient updates at one iteration is equal to the number of elements in the training set; therefore, it is necessary to pay special attention to the number of iterations and to verify the convergence of the learning process. It is recommended to use the ReLU activation function for faster convergence of the learning process and various regularization methods to prevent overfitting. All these recommendations were applied in the employed

IOP Conf. Series: Materials Science and Engineering **714** (2020) 012024 doi:10.1088/1757-899X/714/1/012024

convolutional neural network, which gives hope that a good result can be achieved.

The model parameters and the optimal network architecture was selected on a small training sample when recognizing words outside this sample.

The optimal parameters of the neural network are 10000 for the number of learning iterations, 0.0003 for the learning speed and 600 for the size of the pack.

4. Results

4.1 Own speaker recognition without noise

The following experiment was carried out. Samples of each speaker were divided into 2 parts. In the first part, a convolutional neural network was trained, and the second part was used as a test set. The results of these recognitions are shown in the first line in the Figure 1. The columns show the number of samples for each word in the training set. The rows indicate the noise level in recognized samples which would be described in the following subsections.

Noise level	Number of samples in training set								
	1	2	5	7	9	15	21	25	29
w/o noise	29.5	13.4	3.8	1.8	1.8	0.9	0.5	0.4	1.1
SNR = 6	27.5	14.5	4.1	2.6	2.0	1.6	0.6	0.9	1.1
SNR = 4	25.6	13.8	5.1	3.2	2.5	1.7	0.8	1.1	1.1
SNR = 2	27.3	15.5	5.2	3.7	2.5	1.8	0.9	1.0	1.7
SNR = 1	32.2	16.1	5.8	4.3	3.2	1.9	1.5	1.1	2.2
SNR = 0.5	35.6	18.8	7.0	5.1	4.8	2.8	1.7	2.1	2.2

Figure 1. The average results of word recognition by a convolutional neural network under different noise levels when learning on the same speaker, the table shows the percentage of incorrect recognition.

One could see that it is enough to study on 7 samples for the error rate to fall below 2%, on 15 samples to falls below 1% and on 21 samples for 0.5% of errors. These results show what error values should be sought when recognizing samples of other speakers to exclude the effect of speaker dependence.

4.2 Other speakers recognition without noise

After that, recognition was carried out using samples of a single speaker. This means that the neural network was trained on the samples of one speaker and then the samples of all speakers were recognized on this neural network. When recognizing words from a training set, an average of 0.0% of errors is obtained, and for recognizing words from a test set, 5.9% of errors.

Further, recognition was carried out using samples of three speakers. When recognizing words from a training set, on average, 0.0% of errors are obtained, and for recognizing words from a test set, 1.7% of errors.

For recognition using a larger number of speakers, several training sets were constructed. Each set consists of samples of 7 different speakers. Further, on neural networks trained on compiled sets, a recognition was made on training and test samples. There is no errors when recognizing words from the training set, and on average, 0.6% of errors are obtained for recognizing words from the test set.

The Table 1 shows the total results of recognition of samples without noise when learning on a various number of speakers. From the results we can conclude that the percentage of errors steadily decreases with an increase in the number of speakers in the training set. A greater number of speakers allows you to get rid of speaker dependence, which positively affects the quality of recognition.

Table 1. Words recognition results on training sets containing a different number of speakers for the case of samples without noise.

Number of speakers	Percentage of errors
1 speaker	5.9%
2 speakers	2.4%
3 speakers	1.7%
4 speakers	1.2%
5 speakers	1.0%
6 speakers	0.8%
7 speakers	0.6%

4.3 Own speaker recognition with noise

This subsection presents the results of recognition of samples with noise. The noise was modeled as follows. For each sample, unique pieces of noise were selected from the noise record of the cockpit of a Boeing aircraft with an average volume of more than 80 dB. Also, the noise volume was normalized so that the signal-to-noise ratio was equal to some fixed value.

An experiment was conducted to recognize samples of that speaker, samples of which were used as a training set. As well as in the case of noise-free samples, samples of each speaker were divided into 2 parts: the convolutional neural network was trained on the first part, and the second part was used as a test set. The results of these recognitions are shown in the Figure 1. The columns show the number of entries for each word contained in the training set. The columns indicate the level of added noise. The results are averaged over all speakers.

It can be seen that the results monotonically deteriorate with increasing noise level, while the magnitude of the deterioration is negligible. For example, when using 15 samples in the training set, in the absence of noise, 0.9% of errors are obtained, and with a signal-to-noise ratio of 1, 1.9% of errors are obtained. The similar values for training on 9 samples are 1.8 and 3.2%, respectively. These results show what error values should be sought when recognizing samples with the noise of other speakers to eliminate the effect of speaker dependence.

4.4 Other speakers recognition with noise

After that, the usual recognition was carried out by one speaker with one additional variant of noise. Hereinafter, the noise volume was normalized so that the signal-to-noise ratio was 4. To increase the training set and to improve the learning process, various unique noise variants can be added to samples.

2 variants of the experiment were conducted. In the first variant, training process was conducted for the samples with noise, and samples without noise were recognized. The average error for the recognition of the speaker's samples on which training was conducted is 0.3%, and the average error for the samples of other speakers is 11.8%. In the second variant, both training and recognition were carried out on samples with noise. There were no errors on the samples on which the training was conducted, and the average error for the samples of other speakers was 9.7%. It can be seen from these results that for both recognition options, the error is higher than when using samples without noise. An experiment was also conducted in which training was conducted on samples without noise, and recognition on samples with noise. For this option, the number of errors was several times greater than for the previous ones, so this option will not be used in the future.

Then a similar experiment was carried out, but with the difference that for training one used samples with seven different added noise variants. For the first experiment, the average error in the recognition of the speaker's samples on which training was conducted is equal to 0.0%, and the average error for the samples of other speakers is 8.5%. For the second experiment, there were no errors on the samples on which the training was conducted, and the average error for the samples of other speakers was 8.2%. As can be seen from the results, the use of several noise variants significantly improves the quality of recognition.

After that, the neural network was trained on 3 speakers in noise conditions and subsequent recognition of samples. The first experiment involves the use of one variant of noise in training process, and the second experiment - 3 variants of noise. In both cases, the average error when recognizing the speaker's samples on which training was conducted is equal to 0.0%. For the case of samples recognition of other speakers, the error is 2.8% for the first experiment and 2.5% for the second.

From the results obtained in this and previous experiments, it can be concluded that the use of additional noise options added on the same source samples helps to improve the training process of the neural network and, as a result, the quality of recognition of speech commands. That is, practice shows that the more noise options are used, the better the results. The only limitation is the computing power of the computer used to train the neural network.

For the experiment on samples with added noise during training on 7 speakers, 3 variants of noise were used. When recognizing words from a training set, the average error is 0.0%, and for recognizing words from a test set, an average of 1.1% errors were obtained.

The Table 2 shows the total results of recognition of samples with noise when learning on a various number of speakers and a various amount of added noises.

Table 2. The total recognition results on the training set and on the test set when learning on a various number of speakers for cases of samples without noise and with noise, the table shows the percentage of incorrect recognition.

Number of speakers	Noise in training set	Number of noises	Percent of errors
1 speaker	without noise	1	11.8%
		7	8.5%
	with noise	1	9.7%
		7	8.2%
3 speakers	with noise	1	2.8%
		3	2.5%
7 speakers	with noise	1	1.2%
		3	1.1%

From the results we can conclude, similar to the conclusion in the previous subsection, that the percentage of errors steadily decreases for the same reasons with an increase in the number of speakers in the training base. It can also be concluded that an increase in the amount of added noises leads to a decrease in the number of errors, but this effect from an increase in the training set is less pronounced. A large number of noises in the training set reduces the effect of overfitting for a certain type of noise, which reduces the number of errors in the recognition of samples with other noise. In other words, "noise independence" is similar to speaker independence and is achieved by increasing the number of different noises in the training sample.

Both the results of experiments with noise and the results without noise turned out to be the best when using convolutional neural networks of deep learning. The disadvantages of this algorithm include the great computational complexity of network training. At the same time, recognition itself occurs almost instantly.

5. Conclusion

In this work, one proposed the convolutional neural network architecture, which showed good recognition results in solving several different cases with and without noise.

The first case was a recognition of samples of a speaker whose samples were also used in training process. The results showed that it is possible to achieve a recognition error of 2.0% by using 7 samples of each word in training set and a recognition error of 0.5% for 20 samples of each word.

The second case was to train on samples of one or more speakers and then to recognize sample of someone else's speaker. For this configuration, a share of errors is equal to 5.9% for a training sample of one speaker, 1.7% for a sample of 3 speakers and 0.6% for 7 speakers. The last result is very good, since the result of speaker-independent recognition comes close to the result of speaker-dependent recognition.

The third case was about testing the efficiency of convolutional neural networks during training and recognition of samples with noise. Here, unique parts of the cockpit noise were added to all samples. When recognizing samples of own speaker, the number of errors increased by about 2 times when using a signal-to-noise ratio of 1. A stable, but not very large increase in the number of errors with increasing noise level was also shown. When recognizing samples of other speaker, a share of errors was equal to 9.7% for learning from samples of one speaker with one added noise, and if 7 different noise samples were added to the training set, the error was reduced to 8.2%. For 3 speakers, the error rate was 2.8% for one noise sample and 2.5% for 3 noise samples. For 7 speakers these values were 1.2 and 1.1% respectively, which is an extremely low value for recognition under noise conditions.

Acknowledgements

This work was supported by Russian Foundation for Basic Research (RFBR), project 18-08-01142-a.

References

[1] Evdokimenkov V, Kim R, Krasil'shchikov M and Sebryakov G 2015 *Journal of Computer and Systems Sciences International* **4** 609–20
[2] Polyak B and Khlebnikov M 2017 *Automation and Remote Control* **3** 490–506
[3] Kolokolov A 2006 *Problemy upravleniya* **3** 13–8
[4] Rabiner L 1993 *Fundamentals of Speech Recognition*
[5] Benesty J, Sondhi M and Huang Y 2007 *Springer Handbook on Speech Processing and Speech Communication* (Heidelberg: Springer-Verlag Berlin)
[6] Schmidt-Nielsen A, Marsh E, Tardeli J, Gatewood P, Kreamer E, Tremain T, Cieri C and Wright J 2000 *Speech in Noisy Environments (SPINE) Evaluation Audio* (Philadelphia, PA: Linguistic Data Consortium)
[7] Kolokolov A and Lublinsky I 2015 *Automation and Remote Control* **10** 144–51
[8] Savchenko L 2014 *Informatsionno-upravlyayushchiye sistemy* **1** 23–31
[9] Korsun O and Poliev A 2016 *Journal of Computer and Systems Sciences International* **4** 609–18
[10] Oppenheim A, Buck J and Shafer R 2001 *Discrete-Time Signal Processing*
[11] TensorFlow: An open source machine learning framework for everyone 2018 https://www.tensorflow.org
[12] TFLearn: Deep learning library featuring a higher-level API for TensorFlow 2018 http://tflearn.org
[13] Dieleman S, van den Oord A, Korshunova I, Burms J, Degrave J, Pigou L and Buteneers P 2015 http://benanne.github.io/2015/03/17/plankton.html
[14] Truong T, Nguyen V and Tran M 2018 *ICPRAM* 675–82
[15] CIFAR-10 and CIFAR-100 datasets 2018 https://www.cs.toronto.edu/~kriz/cifar.html
[16] Andrew L 2017 https://beamandrew.github.io/deeplearning/2017/06/04/deep_learning_works.html

Model-Based Optimization in Induction Heating of Thin-Wall Shells

N G Rogachev[1] and G N Rogachev[1]

[1] Institute of Automation and Information Technology, Samara State Technical University, Samara, Russia

E-mail: grogachev@mail.ru

Received xxxxxx
Accepted for publication xxxxxx
Published xxxxxx

Abstract

A mathematical model of induction heating process of paramagnetic thin-wall shells was considered. Electromagnetic processes in the system “inductor heated shell-temperature equalization devices” are modelled by the method of coupled circuits, thermal processes — by the differential-difference method. The model is implemented in MATLAB and is used to optimize the heating process based on partial criteria that are formulated in fuzzy form.

Keywords: fuzzy optimization, control, modeling, induction heating, thin-walled shell

1. Introduction

Induction heating has many advantages: low inertia, energy saving, small size, etc. But it also has disadvantages, the main of which is the irregularity distribution of temperature over the heated object. Therefore, we need to find solution that will prevent this phenomenon. This work is devoted to the problems of model-oriented design and control in induction heating of thin-wall shells.

The aim of our study was to develop a control system for the induction heating of paramagnetic thin-walled shells to specified conditions. The system should be near optimal for a set of partial criteria formulated in fuzzy form. In our study we find program for a digital controller of a hybrid continuous-discrete control system. The continuous part is the system "inductor — heated shell — additional coils" as non-linear MIMO object. The discrete part is the digital controller. The developed digital controller must ensure that the most general requirements for controlled processes are met, including those that are not clearly defined.

The considered unit is shown in figure 1. The shell is placed inside the induction heater, consisting of an inductor powered by an AC source, and thermal insulation. The heating temperature is controlled by thermocouples at three points spaced along the length of the heated body (signals y_1, y_2, y_3). The temperature field at the final stage of the process should be fairly uniform $y_1, y_2, y_3 \in [y_{SP} - \varepsilon, y_{SP} + \varepsilon]$. This requirement is ensured by three control actions u_1, u_2, u_3. The main control channel is the power source of the inductor $u_1 \in [u_{\min}, u_{\max}]$. This channel allows you to influence the average temperature of the shell, but it is not able to significantly change the law of temperature distribution along its length. To ensure the necessary uniformity of heating, the inductor is equipped with additional coils as a device for uniformity the temperature along the length of the shell. When opening $(u_i = 0,\ i = 2,3)$ the coils do not affect the heating process. When closing the coil $(u_i = 1)$, it causes a counter EMF. This leads to a local decrease in heating power in the coil local area, which can help equalize the temperature in the heated product. As it need temperature equalization devices are activated together or alternately, operating simultaneously with the induction heater.

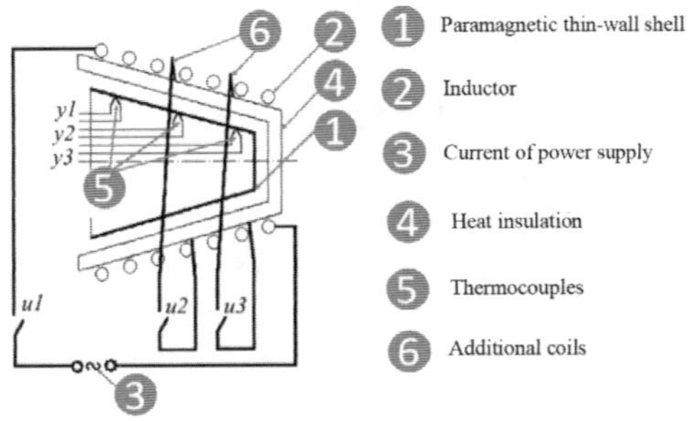

Figure 1. Induction heating unit.

In many existing works, the study of complex multioperational control systems is based on traditional approaches. The mutual influence of both individual parts of the process and the control subsystems is ignored. The synthesis of stage control algorithms is carried out without taking into account the influence of previous stages on the quality of subsequent ones.

The novelty of the proposed approach lies in the organization of control algorithms on digital controllers in the form of "condition-action" type rule systems. The task of finding rule systems is formulated as a fuzzy optimization problem. An important point is also to take into account when determining the operation algorithms of individual control subsystems not only the requirements for the current stage, but also the possible consequences for subsequent ones.

To solve the formulated problem, it is necessary to have a model for describing the induction heating process with the necessary accuracy. The induction heating process can be divided into two parts: energy transfer from an inductor fed by alternating current of a certain frequency to a heated shellp by means of an electromagnetic field; temperature changing of a shell from initial to required by technology.

2. Modeling of the Inuction Heating Processes

2.1 Thermal model

Thermal process in thin shells can be described by the one-dimensional heat equation

$$\frac{\partial T(x,t)}{\partial t}=\left(\frac{\partial^2 T(x,t)}{\partial x^2}+\frac{\Gamma}{x}\frac{\partial T(x,t)}{\partial x}\right)+P(x,t) \qquad (1)$$

with boundary conditions

$$T(x,0)=T_0(x);\ \frac{\partial T(X_1,t)}{\partial x}=q_{X_1}(T);\ \frac{\partial T(X_2,t)}{\partial x}=q_{X_2}(T),$$

where $T(x,t)$ — the temperature field of the heated body, depending on time t and spatial coordinates $x\in[X_1,X_2]$, $\Gamma\in\{1,2\}$ is the form factor of the heated body, $P(x,t)$ is a power of heat sources, Stefan-Boltzmann border loss

$$q_{X_1}(T)=k\left(T^{*4}(X_1,t)-T_{cp}^{*4}\right),$$

$$q_{X_2}(T)=k\left(T^{*4}(X_2,t)-T_{cp}^{*4}\right),$$

where k depending on the effective degree of blackness of the heated material and the Stefan-Boltzmann constant, T^* and T^*_{cp} are the absolute temperature of the product and environment.

Through discretization by the well-known differential-difference method [1], the thermal process in a thin shell is then described in the form of a state-space model. However, the use of standard programs to optimize induction heating processes seems to be impractical due to their low efficiency under these conditions. To take into account the specifics of the processes and tasks under consideration, the authors developed their own subroutines.

2.2 Electromagnetic model

In [2], the method for calculating electromagnetic fields in non-magnetic and ferromagnetic bodies using coupled circuits was chosen as the base for modeling electromagnetic processes in induction heaters. The calculation of fields in non-magnetic bodies according to [2] is reduced to determining the current $\dot I_{1N}$ in the inductor and currents $\dot I_{1i}$, $i=1,\ 2,\ ...,\ N-1$ in $N-1$ elementary solenoids into the body. To find the currents, a system of equations of the form

$$\left[\dot Z_1\right]\left[\dot I_1\right]=\left[\dot U\right], \qquad (2)$$

obtained by applying the 2-nd Kirchhoff law to each solenoid. In (2) the total resistance $\left[\dot Z_1\right]=\left[R_1\right]+j\left[X_{11}\right]$, where $R_{1ij}=\begin{cases}R_{1i},\ i=k\\ 0,\ i\neq k\end{cases}$, R_{1i} is the active resistance of the i-th solenoid, $X_{1ij}=\begin{cases}\omega L_i,\ i=j\\ \omega M_{ik},\ i\neq j\end{cases}$, ω is the angular frequency, L_i — inductance of the i-th solenoid, M_{ik} — mutual inductance of the i-th and k-th solenoids; $\left[\dot U\right]=\left[0,\ 0,\ ...,\ \dot U\right]^T$.

The solution of (2) determines the value of the currents in the inductor and in the solenoids, which represent the components of the body's electromagnetic system. Solenoid currents are used below to determine the power of heat sources $P(x,t)$ in (1),

$$P_{nj}=I_{nj}^2\,Re(\dot Z_{nj}). \qquad (3)$$

For the numerical simulation of electromagnetic processes of induction heating, the authors developed special subroutines, which, together with subprograms for calculating thermal processes, comprise a digital model of induction heating of paramagnetic thin-walled shells. The model was used for fuzzy optimization of a multi-stage heating process.

3. Formulation and Solution of the Fuzzy Optimal Control Problem

Consideration of problems of synthesis of systems of rules for the operation of digital controllers of hybrid continuous-discrete control systems has been the subject of a significant number of studies. In [3], it was noted that in relation to a number of applied problems of synthesis (including optimal) of computer control systems, it is possible either to carry out preliminary finite-dimensional parameterization of some elements of the production system that determines the operation algorithm of the desired controller, or to determine the set of elementary functions from which to construct decision. This allows us to solve such problems numerically as finite-dimensional optimization problems. The traditional approach to solving such problems involves formulating them as mathematical programming problems of the form

$$\min_{z\in Z} f(z),\ \ g_i(z)\geq 0, i=1,2,...,m, \qquad (4)$$

where $z\in Z^*$ is the desired solution, $f(z)$ is the objective function, $g_i(z)$ are the constraint.

Problem (4) can be reformulated so that the boundaries that separate acceptable solutions from unacceptable become blurred, and the degrees of acceptability of solutions are represented by fuzzy numbers. For this, the objective function and limitations must be understood in a fuzzy sense [4]. When using the fuzzy logic notation, the fuzzy version of problem (4) can be written as follows:

$$\widetilde{\min_{z\in Z}} f(z),\ \ g_i(z)\tilde{\geq} 0, i=1,2,...,m, \qquad (5)$$

where the wavy line "~" is a symbol of fuzzy operation. Let membership functions $\mu_0(.)$, $\mu_i(.)$, $i=1, 2, \dots, m$ represent degrees of accomplishment of a goal and limitations. The solution of the optimization problem (5) should satisfy, as far as possible, both the goal and the constraints, i.e. maximize the minimum of values $\mu_i(.), i=0, 1, \dots, m$. So problem (5) of fuzzy mathematical programming can be transformed [5] into a problem $\max_{z\in Z} C(z)$, where $C(z)$ represents the global degree of satisfaction z with a goal and limitations: $C(z)=\min\{v_0, v_1, ..., v_m\}$. Finally, a fuzzy problem (5) takes the form

$$\max_{z\in Z^*}\min_i(v_i),\ i=0,1,...,m, \qquad (6)$$

$v_0=\mu_0(f(z))$, $v_i=\mu_i(g_i(z))$, $i=1,2,...,m$, membership functions $\mu_0(.)$, $\mu_i(.)$, $i=1, 2, \dots, m$ represent degrees of accomplishment of the goal and constraints, respectively. Further, the problem (6) is solved by numerical methods.

In this work the control of paramagnetic thin-walled shells induction heating is considered. The control task with a fuzzy purpose and limitations is formulated as follows. It is necessary in the time $t_F \to \min$ to ensure the transfer of the model of the control object (1) – (4) to the area $\max_{x\in[X_1,X_2]}|T_F(x)|\leq\varepsilon$, $T_F(x)=(T(x,t_F)-T_{SP})$ under the additional condition $\max_{t\in[0,t_F]}(T_D)\leq 2\varepsilon$, where $T_D=\max_{x\in[X_1,X_2]}T(x,t)-\min_{x\in[X_1,X_2]}T(x,t)$. Membership functions of fuzzy sets are determined programmatically as S-shaped membership functions (for t_F and T_D) and triangular-shaped membership function for T_F :

$$\mu_0(t_F)=\begin{cases}1, & t_F\leq 4\\ 1-(t_F-4)^2/32, & 4\leq t_F\leq 8\\ (t_F-4)^2/32, & 8\leq t_F\leq 12\\ 0, & 12\leq t_F\end{cases};$$

$$\mu_1(\rho_1)=\begin{cases}0, & \rho_1\leq 890\\ (\rho_1-890)/50, & 890\leq\rho_1\leq 940\\ (990-\rho_1)/50, & 940\leq\rho_1\leq 990\\ 0, & 990\leq\rho_1\end{cases};$$

$$\rho_1=\max_{x\in[X_1,X_2]}|T_F(x)|;$$

$$\mu_2(\rho_2)=\begin{cases}1, & \rho_2\leq 100\\ 1-(\rho_2-100)^2/450, & 100\leq\rho_2\leq 115\\ (\rho_2-130)^2/450, & 115\leq\rho_2\leq 130\\ 0, & 130\leq\rho_2\end{cases};$$

$$\rho_2=\max_{t\in[0,t_F]}(T_D).$$

We will find the fuzzy solution from relation (6) for a system of rules for digital program controller with a unit operating frequency (ones per model time unit) in the following form "if $t\in[i,i+1)$ then $u_1(t)=u_{1i}$, $u_2(t)=u_{2i}$, $u_3(t)=u_{3i}$", $i=0, 1, 2, \dots, t_F/t_S-1$ where $t_F=6$, $t_S=0.6$, $u_1(t)$, $u_2(t)$, $u_3(t)$ are respectively the inductor power control signal $u_1(t)\in[u_{min},u_{max}]$ and control signals $u_2(t)\in\{0,1\}, u_3(t)\in\{0,1\}$ of additional coils as temperature equalization devices along the shell length.

The problem of determining the fuzzy optimal control is reduced to the problem of integer programming. The parameter $z_i(t_j)$, $i=1,2,3$, $j=0, 1, 2, \dots, t_F/t_S-1$ at each of the 10 intervals of control constancy is the number of one of four possible options for the distribution of volumetric power density of internal heat sources. Option $z_i=1$ corresponds to a combination $u_1=u_{max}$, $u_2=0$, $u_3=0$, option $z_i=2$ — to a combination $u_1=u_{max}$, $u_2=1$, $u_3=0$, option $z_i=3$ — to a combination of $u_1=u_{max}$, $u_2=0$, $u_3=1$, option $z_i=4$ — to a combination of $u_1=u_{max}$, $u_2=1$, $u_3=1$. These combinations lead to the following four options for power distribution along the length of the shell (figure 2). By brute force method we find one of $4^{10} = 1048576$ options. This mode of the form $z = [1,1,1,1,1,1,1,2,3,1]$ ensures the quality of heating at a level $\min(\mu_i(.)) = 0.8750,\ i = 1,\ 2, 3.$

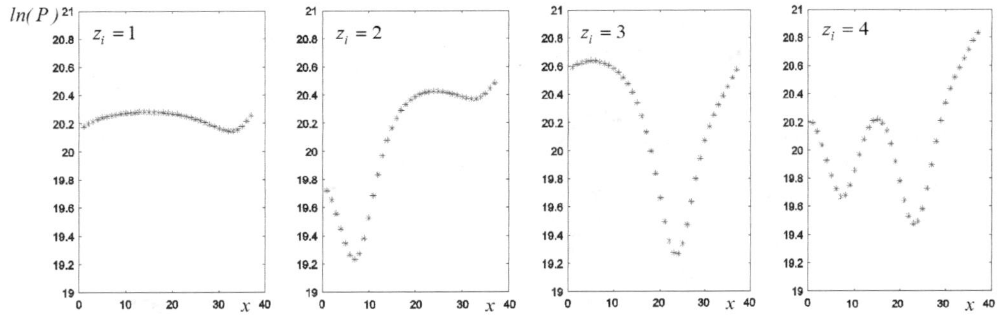

Figure 2. Four possible distributions of volumetric power density of internal heat sources.

The results are presented in figures 3, 4 where $y_1 = T(X_1),\ y_2 = T((X_1 + X_2)/2),\ y_3 = T(X_2),\ t_F = 6,\ t_S = 0.6$. The degree of satisfaction of the obtained solution to the given conditions is 0.875.

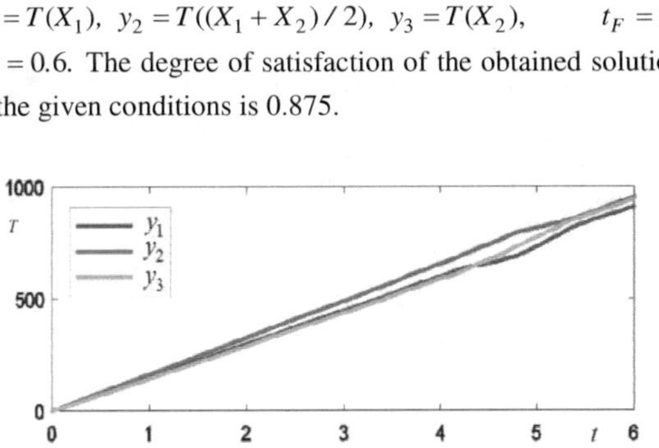

Figure 3. Temperature in a fuzzy optimal problem.

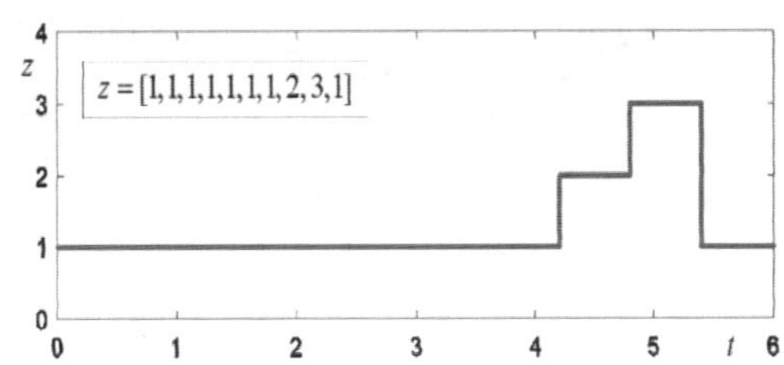

Figure 4. Control actions in a fuzzy optimal problem.

4. Conclusion

This paper demonstrates the possibility of using fuzzy logic in the synthesis of regulators under the conditions of fuzzy formulated requirements for the intermediate and final state of a nonlinear, multidimensional object in the form of a multiply connected process chain. This approach seems very promising. On its basis it is possible to automate the entire process of designing control systems from setting the task to modeling their work, including software testing, creating technical devices that independently generate the algorithms of their control systems and, if necessary, modernizing them.

Acknowledgments

The reported study was funded by RFBR, project number 19-38-90061.

References

[1] Rapoport E Ya 2005 *Analysis and synthesis of automatic control systems with distributed parameters* (Moscow: Higher School)
[2] Nemkov V S and Demidovich V B 1988 *Theory and calculation of induction heating devices* (Leningrad: Energoatomizdat)
[3] Rogachev G N 2006 *Bulletin of Samara State Technical University. Techn. Nauki* **41** 43–7
[4] Bellman R E and Zadeh L A 1970 *Environment Management Science* **17** 4 141–64
[5] Carlsson C, Fullér R and Giove S 1998 *Journal of Operations Research, Statistics and Computer Science* **38** 17–24

IOP Conf. Series: Materials Science and Engineering **714** (2020) 012026 doi:10.1088/1757-899X/714/1/012026

Engineering Software for a Mobile Robot Motion Control System

Igor Ryadchikov[1], Evgeny Nikulchev[2], Alexander Gusev[1], Semyon Sechenev[1] and Alexey Prutskiy[1]

[1] Laboratory of Robotics and Mechatronics, Kuban State University, Krasnodar, Russian Federation
[2] Department of Control and Modelling Systems, MIREA – Russian Technological University, Moscow, Russian Federation

E-mail: alexandrgsv@gmail.com

Received xxxxxx
Accepted for publication xxxxxx
Published xxxxxx

Abstract

The paper centres round the problem of the engineering a motion control system for a mobile robot based on the effective selection of software components with respect to the numerical criterion proposed by the authors. The data for the selection process comes out the reproducible experiments with the sets of alternative components in a Gazebo virtual infrastructure simulating the real robot operating conditions. The genetic algorithm is used to reduce the number of experiments with unpromising sets of software components. The methodology proposed by the authors is applied to the real task of engineering a motion control system for a non-anthropomorphic mobile robot. The virtual infrastructure and genetic algorithm parameters are provided as well as the physical model of the robot for that task. To calculate the integral quality criterion proposed in the paper, 4 partial quality criteria were measured in the experiments with different software components. The motion process of the physical robot with the selected software components is shown.

Keywords: software engineering, quality of software, mobile robot, control system, control moment gyroscope, genetic algorithm

1. Introduction

The effective choice of components for a motion control system of a mobile robot is becoming an important problem of software engineering as the number of components for popular robotics frameworks is increasing [1]. For example, for the Robotic Operating System (ROS) framework there are now thousands of components available. Some of them provide the similar functionality. The interaction of the components in a software stack may lead to an unexpected behavior [2], for example if one of the components blocks the others from accessing a common resource (database, hardware driver, LAN, etc.). Thus, it is important to choose the software components with respect to the quality of their joint functioning in a stack. When developing a real-time motion control system for a mobile robot it is also important to assess the software components in the conditions most similar to the real robot operating conditions. To ensure that the experimental stand for reproducible experiments should be designed. To guarantee the reproducibility of the experiments and to spare the mechanical parts of a real robot it is appropriate to implement the experimental stand in the virtual infrastructure, simulating the operating conditions of the robot using the simulation software like Gazebo [3]. Evolutionary algorithms [4], above all the genetic algorithm [5], can be used to reduce the number of experiments with unpromising software components.

The paper consists of 5 sections: Introduction, Formulation of the problem, Methodology, Solution and Concluding remarks.

2. Formulation of the problem

Let us consider the set of functional features (see table 1) which should be implemented in the motion control system of the robot. For each feature there is a set of alternative software components capable of implementing that feature.

Table 1. List of functional features and components.

Name of the functional feature	Alternative components implementing that functional feature	Description
Servo	Dynamixel mx106 driver Rd50 driver Rd60 driver	Interaction of the control system with the servomotor of a mobile robot
Control	Speedgoat+stm32 Raspberry(Python) Stm32 Stm32+Raspberry	Control of the robot movement with a software module implemented for the chosen architecture of the computer
Gyro	Mpu9250 Gyrolab GL109 Gyrolab GLVG 203	Getting information from a gyroscope

Thus, there are 36 subsets of software components (stacks) in which for every functional feature there is at least one implementing component. The stacks are to be evaluated in the virtual infrastructure, simulating the operating conditions of the robot (see figure 1). Let us denote the stacks as $s^j, j = \overline{1,36}$.

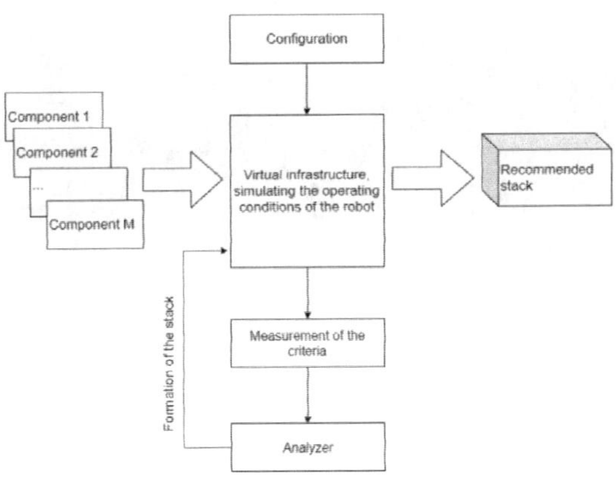

Figure 1. The effective choice of software components.

Table 2 shows the partial quality criteria used to estimate the quality of functioning of the robot motion control system.

Table 2. Partial quality criteria.

Notation	Criterion	Unit
r_1	Response time of the control system	μs
r_2	Mass of the robot including the additional equipment to run the selected software components	kg
r_3	Total mass of additional batteries to ensure autonomous operation of the mobile robot when using the selected software components	kg
r_4	The sum of the squares of the deviations of the robot trajectory from the calculated trajectory for all experimental samples	m^2

Let us define the integral quality criterion of a stack s^j as:

$$\Psi\left(s^j\right) = \sum_{\xi=1}^{4} w_\xi \tilde{r}_\xi^j, \qquad (1)$$

where $\tilde{r}_\xi^j$ are the normalized values of partial quality criteria r_ξ^j; $\xi = \overline{1, 4}$; w_ξ are the weights of the partial criteria: $w_1 = 0.18$; $w_2 = w_3 = 0.16$; $w_4 = 0.5$.

The virtual infrastructure, simulating the operating conditions of the robot is made up with the Gazebo robotics simulator. The configuration of the virtual infrastructure is presented in table 3.

Table 3. Virtual infrastructure configuration.

No	Parameter	Value
1	Physics type	ode
2	Solver type	World
3	Max step size	0.001
4	Gravity	[0.0 0.0 –9.81]
5	Surface contact ode max_vel	10
6	Surface contact ode min_depth	0.001
7	Collision geometry sphere radius	0.5
8	Surface contact ode kp	1e15
9	Surface contact ode kd	1e13

The task is to choose the stack s^*, satisfying the following condition:

$$s^* = \underset{s^j, j=\overline{1,36}}{\operatorname{argmin}} \Psi\left(s^j\right).$$

3. Methodology

To organize a reproducible experiment for evaluation the quality of functioning, a physical model of the robot [6, 7] was assembled (see figure 2) and an experimental stand in the virtual infrastructure was created.

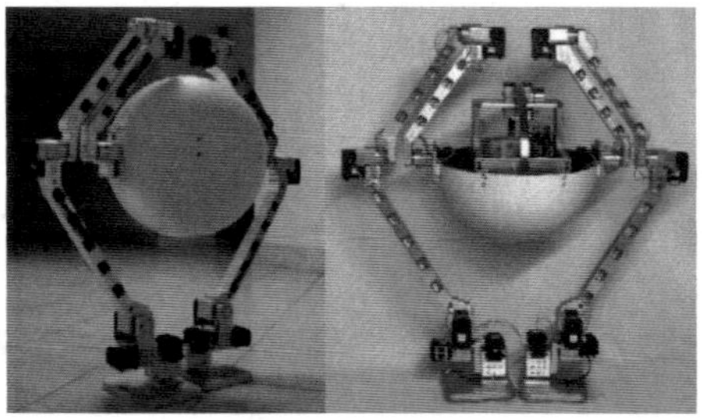

Figure 2. Physical model of the robot.

The automated methodology for selecting an effective set of software components involves using a genetic algorithm to generate and experimentally evaluate stacks of components. Genetic algorithm is a global optimization algorithm based on evolutionary computations. It uses the mechanisms of random selection, combination, and variation of the desired parameters similar to the natural selection.

The motion control process for the experimental stand in the virtual infrastructure and for the physical model of the robot is shown in the figure 3. The genetic search configuration is shown in table 4.

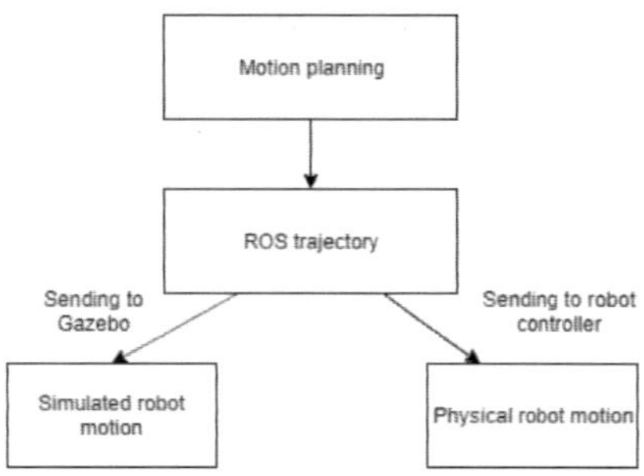

Figure 3. Robot motion control process.

Table 4. Genetic search configuration.

No	Parameter	Value
1	Selection operator	Tournament selection
2	Mutation operator	Extended power mutation
3	Crossover operator	Laplace crossover
4	Probability of mutation	0.01
5	Probability of crossover	0.8
6	Elite count	3
7	Population size	4
8	Maximum generations	100
9	Maximum stall generations	5
10	Function tolerance	0.01

4. Solution

Using *ga* toolkit from the Global Optimization Toolbox of the Matlab R2018a the choice of software components minimizing the integral criterion (1) was made. The graph of evolutionary selection is shown in the figure 4. The motion of the robot is depicted in the figure 5.

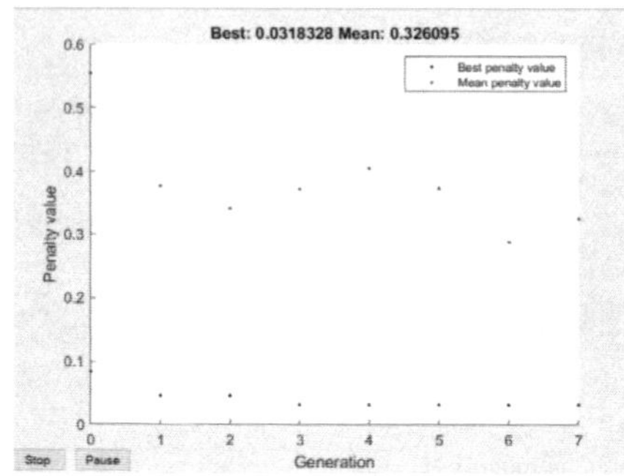

Figure 4. Graph of evolutionary selection of a stack of components. The penalty value refers to the value of integral criterion Ψ for a given stack.

It should be noted that a small "oscillation" in the evolutionary selection graph is explained by the unavoidable measurement noise. However, the genotype of the best solution from generation to generation begins to prevail more and more in the population (this is seen by the decrease in the average value of Ψ) and the identification of the best solution occurs because the number of experiments falling on the best genotype is increasing, which eliminates random factors in assessing this genotype.

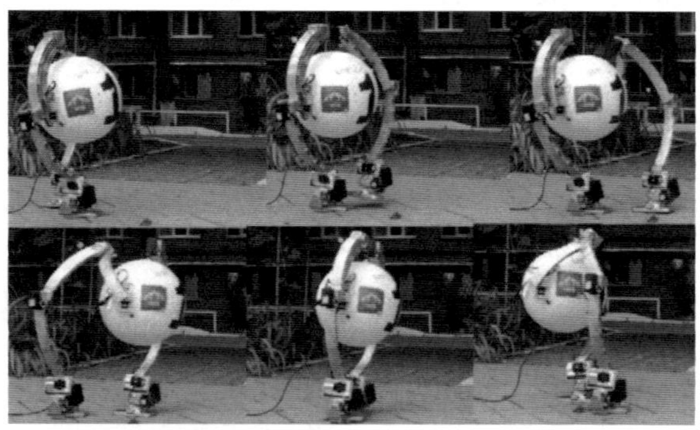

Figure 5. The motion of the robot.

For the considered example, the selected effective solution s^* (genotype — [2 1 2]) is as follows:

- Servo feature is implemented with the "Rd50 Driver" component;
- Control feature is implemented with the "Speedgoat + stm32" component;
- Gyro feature is implemented with the "Gyrolab GL109" component.

5. Concluding remarks

The results of the practical implementation of the methodology for the experimental selection of software components demonstrate the applicability and the effectiveness of the proposed approach in engineering software for motion control systems of mobile robots. Further research will be aimed at improving the methodology in terms of increasing the convergence rate of the evolutionary algorithm and considering larger sets of partial quality criteria.

Acknowledgements

The work was carried out within the framework of the state task of the Ministry of Science and Higher Education, project No. 8.2321.2017/4.6 “Development and adaptation of control systems for compensation of dynamic deflecting effects on mobile objects in a state of dynamic equilibrium”.

References

[1] Romanov A M 2019 A review on control systems hardware and software for robots of various scale and purpose *Russian Technological Journal* **7**(5) 30–46 (doi: 10.32362/2500-316X-2019-7-5-30-46)

[2] Wienke J, Wigand D, Koster N and Wrede S 2018 *Proc Second IEEE International Conference on Robotics Computing (Laguna Hills, CA, USA)* (Hoboken, New Jersey: I.E.E.E. Press) pp 25–32

[3] Koenig N P and Howard A 2004 *Proc. IEEE/RSJ International Conference on Intelligent Robots and Systems (IROS) (Sendai, Japan)* (Hoboken, New Jersey: I.E.E.E. Press) 3 pp 2149–54

[4] Xiaohu Y, Fazhi He, Neng H and Haojun A 2018 *International Journal of Cooperative Information Systems* **27**(1) 1741001

[5] Patle B K, Parhi D R K, Jagadeesh A and Kashyap S K 2018 *Computers & Electrical Engineering* **67** 708

[6] Ryadchikov I, Sechenev S, Sinitsa S.,... and Nikulchev E 2017 *International Journal of Advanced Computer Science and Applications* **8**(9) 29–34

[7] Ryadchikov I., Sechenev S., Nikulchev E … and Vishnykov R 2018 *International Review of Automatic Control*, **11**(4) 160–5

Mathematical Model of Lithium-Ion Battery Cell and Battery (Lib) on its Basis

D I Smagin[1], A A Trofimov[1], K S Napreenko[1] and A R Neveshkina[1]

[1] Moscow Aviation Institute (National Research University), Moscow, Russia

E-mail: 79637587781@yandex.ru

Received xxxxxx
Accepted for publication xxxxxx
Published xxxxxx

Abstract

The article considers a mathematical model of lithium-ion battery cell and battery (LIB) on its basis. The developed mathematical model allows predicting LIB temperature on different parts of its surface during charging and discharging by nominal and maximum currents. The results of the battery discharge process simulation and validation of the mathematical model based on the field experiment results are also presented

Keywords: lithium-ion battery, mathematical model, power supply system, thermal analysis, electric transport

1. Introduction

The use of mineral fuel and energy resources and the depletion of its reserves are one of the most pressing problems in the modern world. The fuel use, which adversely affects the environment, provokes environmental issue. The environmental safety requirements, which are becoming more stringent every year, demand the development and creation of environmentally friendly energy sources that can replace traditional hydrocarbon reserves in the main industries, including transport [1-3].

Currently, one of the most relevant areas of the transport industry development is the creation of electric transport using lithium-ion batteries [4]. Such batteries have a few advantages, among which are high energy density, minimum self-discharge, weight and size indicators, etc. [5–7]. Today LIB applied in the areas of transport sector, public transport [8] (buses, trams and trolleybuses with increased autonomous course), sea and railway transport and aviation perspective aircraft with a high level of electrification [9].

Since power supply system is a fundamental system during development and operation of electric transport then for its full operation is necessary to have enough information about the performance of the components which constitute it.

One of the main factors affecting the work of LIB is heat. The operating temperature of the battery cells affects such important parameters of LIB operation as energy efficiency, reLIBility, service life and, as a result, the life cycle cost [10-12]. For a detailed thermal analysis of such energy sources it is advisable to use a mathematical model.

2. Mathematical model functions

The main functions of the mathematical model LIB are:

- Forecasting the temperature of battery cells under different operating modes;
- Analysis of LIB heat release under load with different currents.

3. Mathematical modelling of lithium-ion battery

The basis of the mathematical model is the heat transfer equations.

Conductive heat transfer is determined by the Fourier`s law of thermal conductivity:

$$q = -K_{nn} \frac{\partial T}{\partial n}, \quad (1)$$

q — heat flow rate per unit area in direction n; K_{nn} — thermal conductivity in direction n; T — temperature; $\frac{\partial T}{\partial n}$ — temperature gradient in direction n.

Convection heat transfer is determined by Newton's law:

$$q = h \cdot \Delta T, \tag{2}$$

q — heat flow rate per unit area between surface and matter; h — convective film coefficient; ΔT—the temperature difference between surface and matter.

Radiation is described by Stefan-Boltzmann law:

$$q = \sigma \cdot \varepsilon \cdot A_i \cdot F_{ij}(T_i^4 - T_j^4), \tag{3}$$

q — heat flow rate from surface i to surface j; σ — «Stefan-Boltzmann constant»; ε — emissivity; Ai — area of surface i; Fij — form factor from surface i to surface j; Ti — absolute temperature of surface i; Tj — absolute temperature of surface j.

The heat conduction equation for a differential volume can be written in the form:

$$\rho c \frac{\partial T}{\partial t} + Q - K\,\nabla^2 T = 0, \tag{4}$$

ρ — density of the substance; c — heat capacity; T — temperature; t — time; Q — the rate of internal heat generation rate per unit volume; K — specific thermal conductivity tensor; $\nabla^2 T$— temperature Laplacian.

The Biot Number and The Fourier Number are used to determine the optimal integration step in the mathematical model.

The Biot Number characterizes the relationship between the temperature difference in the two points of the body at a certain distance and the temperature pressure — the difference between the body surface temperature and the ambient temperature. The Biot Number is determined by the dimensionless ratio:

$$B_i = \frac{h\Delta x}{K}, \tag{5}$$

B_i— The Biot Number; h — average heat transfer coefficient; Δx — average width of the element; K — average thermal conductivity.

The Fourier Number characterizes the ratio between the thermal conditions rate change in the environment and the rate of temperature field change within the system (body):

$$F_0 = \frac{4K\Delta t}{\rho C(\Delta x)^2}, \tag{6}$$

F_0 — The Fourier Number; K — average thermal conductivity; Δt — time frame; ρ — density of the substance; c — heat capacity; Δx — average width of the element.

In case the Biot Number is less than one for the calculation of the integration is used Fourier number. Otherwise, for the calculation of the integration is used determination of this numbers.

As the research objects it were selected battery cells with a nominal capacity of 170 Ah of each one. The main technical characteristics of the cell are presented in table 1.

Since the lithium-ion battery is a set of identical battery cells connected to each other, let us consider the thermal mathematical model for the case of one battery cell, 3D model of which is shown in figure 1.

The model consists of four main elements: housing, positive terminal, negative terminal and core.

Table 1. Battery cell specifications

Parameter		Value
Nominal Capacity		170 Ah
Nominal Voltage		3,2 V
Weight		6,9 kg
Internal Impedance at 1 kHz		0,5 mOm
Dimension	Length	150 mm
	Width	85 mm
	Height	346 mm

Figure 1. Battery cell 3D model.

The battery cell core material is considered as homogeneous [13]. Core characteristics (density, thermal conductivity and heat capacity) correspond to those described in articles [13–15].

The internal heat release of the battery cell depends on the amount of current flowing.

4. Simulation result

In this example, heat exchange with the external environment is due to natural convection, the coefficient of convective heat transfer is 5 W/(m2·°C). Radiation in this task can be neglected, because the temperature on the surface of the battery cell is slightly higher than the ambient temperature and, therefore, the effect of radiation on the heat release process will be insignificant.

The calculated temperature distribution for the case of the battery cell is shown in figure 2, for the case of the battery consisting of 4 cells — in figure 3.

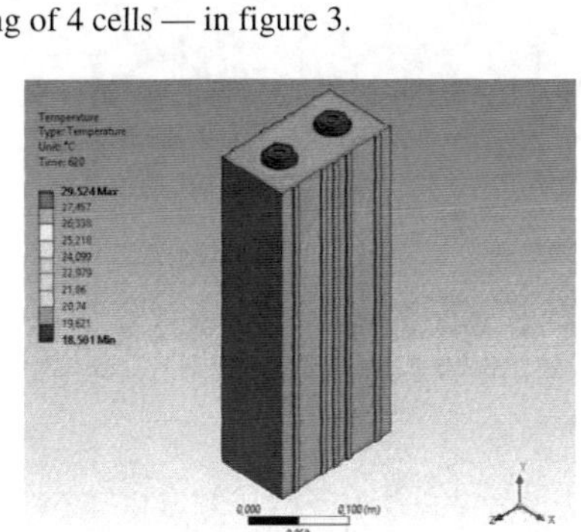

Figure 2. Calculation results for one battery cell.

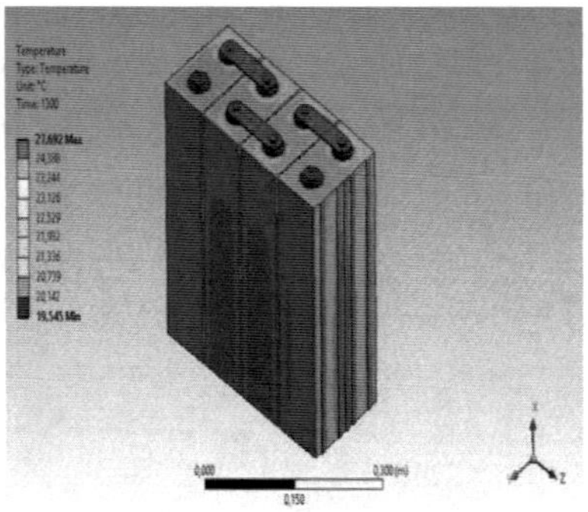

Figure 3. Calculation results for the battery.

5. Testing LIB

The aim of the exercise tests was determination of lithium-ion battery temperature on different parts of its surface during the discharge of nominal and maximum currents.

The test bench is a processing board of information coming from the temperature sensors, which are placed on LIB.

The experiments were conducted at an ambient temperature of 19.5 °C and natural convection. The battery consisting of 4 such cells was assembled for achieving the required discharge current. Sensor readings are recorded every 3 seconds. Sensors were located on the most exposed to heat LIB elements — terminals, cell connectors and on the surface of the body. The schematic arrangement of the sensors is shown in figure 4. It allows obtaining a wide picture of the main LIB sections temperature distribution. Temperatures in the controlled points at the discharge current 1C are shown in figure 5.

DS18B20 digital sensors with ±0.5°C measurement error was used as temperature sensors [16]. Information is exchanged with the master (microcontroller or Arduino board) via the 1-wire bus. LIB temperature was also controlled by thermal imager (figure 6).

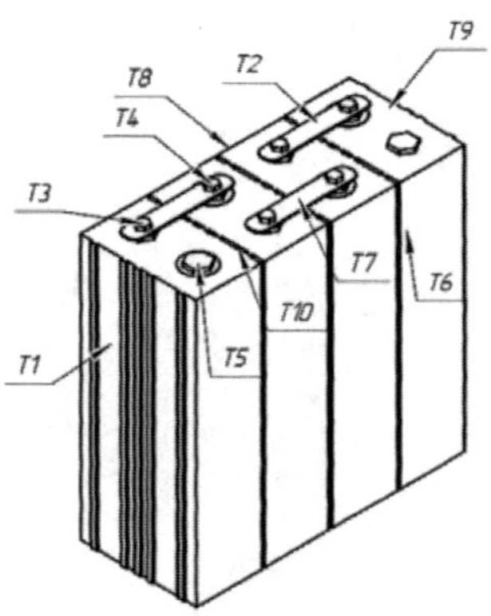

Figure 4. The location of the temperature sensors.

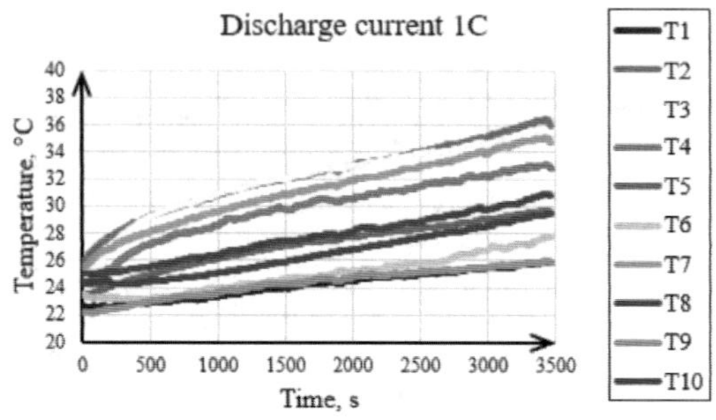

Figure 5. Temperature of LIB elements at discharge current 1C.

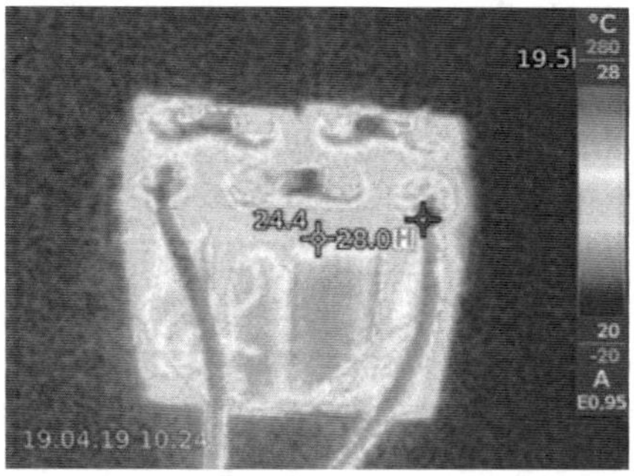

Figure 6. LIB temperature distributions by thermal imager.

It should be noted that the thermal analysis of the LIB results is comparable with the heat release of batteries studied in articles [14, 17].

Validation of thermal mathematical model was carried out based on experiments on the discharge of a lithium-ion battery consisting of 4 cells.

Figure 7 shows a graph of the battery current 1C discharge, which shows the results of the experiment (temperature of the sensor located on the terminal) and mathematical modeling at the same point. Dashed line indicates a range of ±1.5°C from the experimental data.

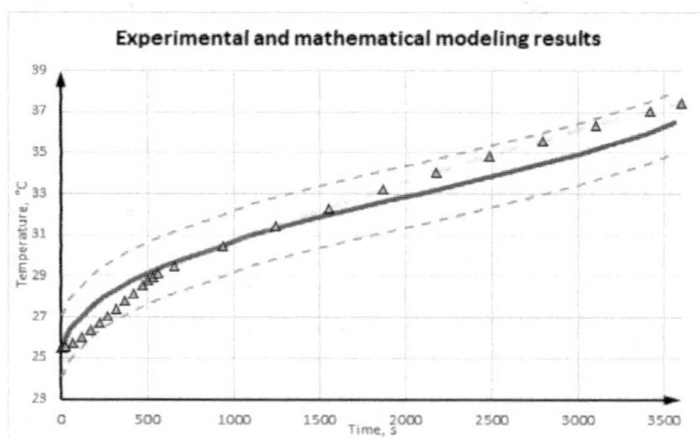

Figure 7. Temperature versus time graph for stationary operation.

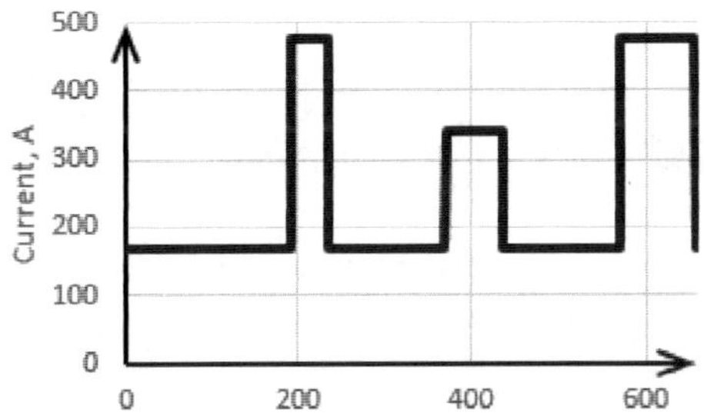

Figure 8. A graph of the discharge current.

For estimating the unsteady process, the discharge current ranged from 1C to 2.8 C, as it is shown in figure 8.

Figure 9 shows a graph illustrating the results of the experiment and mathematical modeling at the same point for the unsteady process. The dashed line indicates the range of ±1°C from the experimental data.

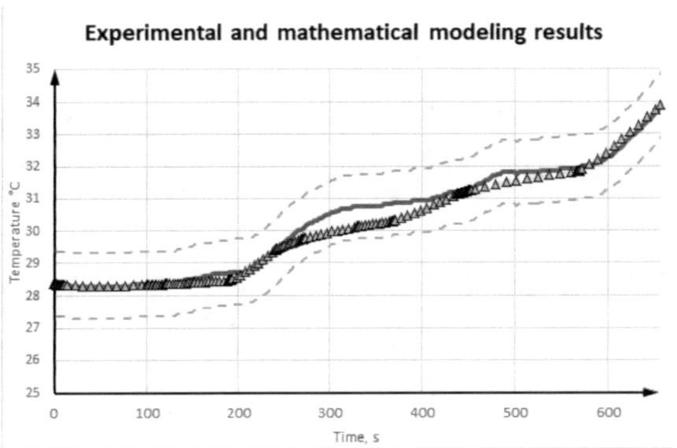

Figure 9. Temperature versus time graph for non-stationary mode.

As can be seen from the graphs shown in figure 7 and 9, the greatest discrepancy between the results of the experiment and mathematical modeling is about 1 °C for a stationary process and does not exceed 0.6 °C for a non-stationary one. This discrepancy results may be caused by the inaccuracy of the temperature sensors and possible minor computational errors associated with the construction of a computational grid. The results of the battery validation allow us to conclude that all the parameters of the battery cell are considered correctly, and the developed mathematical model can be used to analyze the heat release of LIB, consisting of battery cells given number.

6. Conclusion

The developed lithium-ion battery mathematical model showed good convergence of the results with full-scale tests, which will not only predict the heat release of the battery in different operation modes with a high degree of accuracy, but also to use this mathematical model in the future, for example, in the development of a thermostatic system.

References

[1] Boureima F S 2011 *Environmental Assessment of Conventional and Alternative Vehicles and Fuels in a Belgian Context* PhD Thesis (Brussels, Belgium: Vrije Universiteit Brussel)
[2] Van Mierlo J and Maggetto G 2003 *J. Automobile Engineering* **7** 583–93
[3] *Chemical sources of current*: *The manual* ed by Korovin N In and Skundin A M 2003 (Moscow: MPEI Publishing house)
[4] Chen S C, Wan C C and Wang Y Y *2005 Journal of Power Sources* **140** 111–24
[5] Pollet B, Staffell I and Shang J 2012 *Electrochem. Acta* **84** 235–49
[6] Wu B, Yufit V, Marinescu M, Offer G J M – B, Ricardo F and Brandon N P 2013 *J. Power Sources* **243** 544–54
[7] Bystrov Y A, Kudryavtsev N A, Krasnobryzhy V A, Loginova M M and Zhdanov V V 2007 *Underwater research and robotics* **2** 34–7
[8] Klimchuk A K and Temirgaliyev E R 2017 Possibilities of introduction of public electric transport instead of traditional (on the example of St. Petersburg) *Proceedings of the scientific conference with international participation* (St. Petersburg: Federal state Autonomous educational institution of higher education "Peter the great St. Petersburg Polytechnic University")
[9] https://www.aeroflap.com.br/airbus-conclui-que-bateria-de-litio-ion-do-a350-e-confiavel/
[10] Pesaran A A 2002 *J. Power Sci.* **110** 377–82
[11] Rothgang S, Rogge M, Becker J and Sauer D U 2015 *Energies* **8** 6715–37
[12] Shabani B and Andrews J 2015 *Int. J. Hydrogen Energy* **40** 5509–17
[13] Liu X, Zhang Y and Li X 2017 Battery Pack's Structure Design Based on the Heat Flow Field Simulation *2nd International Conference on Industrial Aerodynamics (ICIA 2017)* pp 614–23
[14] Samba A, Omar N, Gualous H, Van den Bossche P, Van Mierlo J and Boubekeur T I 2013 Development of 2D Thermal Battery Model for Lithium-ion Pouch Cells *EVS 27 Barcelona, Spain*
[15] Arunkumar R, Anbumalar S, AvilaPriya F and Poongothai K 2015 *Int. J. of Sci. & Eng. Res.* **6 4** 74–8
[16] http://pdf.datasheetcatalog.com/datasheet/maxim/DS18B20.pdf
[17] Yi J, Koo B and Shin C B 2014 *Energies* **11** 7586–60

IOP Conf. Series: Materials Science and Engineering **714** (2020) 012028 doi:10.1088/1757-899X/714/1/012028

Hybrid Sensor for Determining the Angular Position

A A Tsatsin[1]

[1] Joint Stock Company «Avionica», Moscow, Russia

E-mail: sand_e@mail.ru

Received xxxxxx
Accepted for publication xxxxxx
Published xxxxxx

Abstract

The basic provisions of the approach to the creation of hybrid angular position sensors, the principle of output signal correction are considered.

Keywords: analog angular position sensor, digital angular position sensor, hybrid angular position sensor, correction

1. Introduction

In the development of automatic control systems aircraft inevitably raises the question of the choice position sensors various objects: controls, steering surfaces, etc. Currently, there is a steady trend of transition from analog to digital sensors, while one of the main problems remains the need to ensure the specified accuracy of the sensor.

This article discusses the basic provisions of creating a prototype hybrid angular position sensor, which includes both analog and digital sensor, connected in a certain way both structurally and algorithmically in order to obtain a single output signal.

In aviation systems, analog sensors in their pure form are rarely used, more often they are used in conjunction with an analog-to-digital converter of the output signal. This integration makes it possible to significantly simplify the processing of the output signal in modern digital computing systems.

At the same time, the main disadvantage of analog sensors is a relatively high dependence of the output signal on external influences: temperature, humidity, pressure, etc.

Digital position sensors are significantly less affected by the external environment, but have their own disadvantages:

- incremental sensors have not yet found application in aviation, because in case of loss and subsequent restoration of power supply it is not possible to determine the current object angular position;
- absolute sensors with high accuracy have relatively large dimensions, which is why there are difficulties in the organization of redundancy channels to obtain information about the current position.

2. Basic information

To overcome the current situation with the shortcomings of existing angular position sensors, it is proposed to create a Hybrid Sensor (HS), the block diagram of which is presented in figure 1 [1].

The figure shows that the hybrid position sensor includes the following main elements:

- Digital Sensors (DS);
- Analog Sensor (AS);
- Analog-to-Digital Converter (ADC);
- Controller.

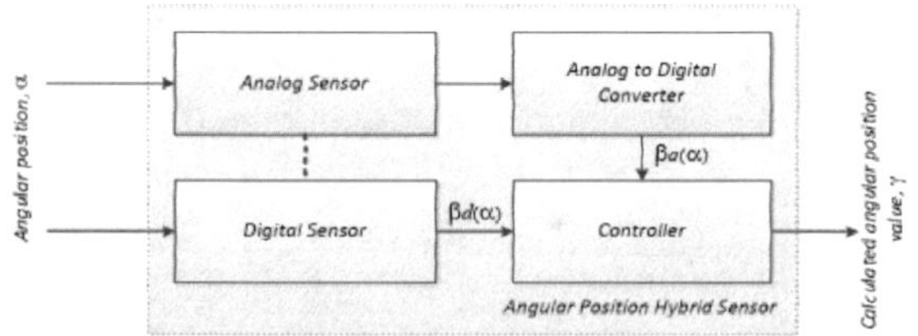

Figure 1. Block diagram of the Hybrid Sensor.

The basic idea of hybridization is to use different measurement principles. In this case, simultaneously with the analog angular position sensor, a digital direct conversion sensor "position-code" is used.

The sensors included in the hybrid have different approaches to measuring the angular position, so it is convenient to model and analyze the output characteristics of the sensors in a rectangular Cartesian coordinate system "specified angular position-output angular position". This allows you to get away from the consideration of physical phenomena occurring in the sensors and the parameters describing them.

At the initial stage of research in the mathematical modeling of the hybrid sensor, we assume the following assumptions:

- no noise of analog sensor and ADC;
- no "rattle" of the digital sensor.

For example, consider a hybrid angular position sensor, which includes a hypothetical digital and analog sensors with output characteristics, respectively, $\beta d(\alpha)$ and $\beta a(\alpha)$ (figure 2). The figure also shows the output characteristic of the angular position Ideal Sensor (IS).

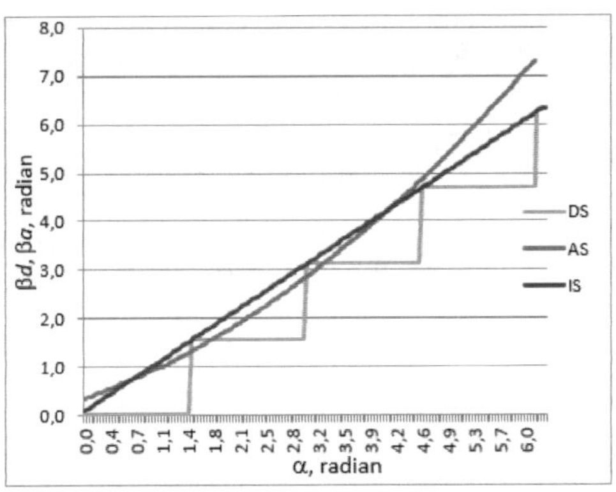

Figure 2. Output characteristics DS, AS and IS.

As can be seen from figure 2, the output characteristic $\beta a(\alpha)$ of a hypothetical Analog Sensor has bad features:

- different degree of curvature in different parts of the characteristic;
- there are offsets characteristics at α=0 and $\alpha=2\pi$;
- the characteristic offsets in α=0 and $\alpha=2\pi$ are not equal.

As a prototype, a mathematical model of a single-track binary Digital Sensor based on the T-sequence T={0011}.

Digital Sensor — binary, has a bit width n_b=2 and therefore has the number of angular positions (quant) N equal to

$$N = 2^{n_b} = 2^2 = 4.$$

For any i-th quants of digital sensor, knowing the current values of the signals βd_i , βa, digital and analog sensors, respectively, it is possible to determine the value of the object current angle γ [2].

$$\gamma = \frac{\beta_a - \beta_{a_i}}{k_i} + \beta_{di}, \quad (1)$$

where

$$k_i = \frac{\beta_{a_{i+1}} - \beta_{a_i}}{\alpha_{d_i}} \quad (2)$$

chord slope coefficient of the output characteristic AS in i-th quants,

here βai — the value of the analog sensor signal to the switching instants of the quants; $\alpha_{di} = \beta_{di+1}$ — angular width of the i-th quants digital sensor.

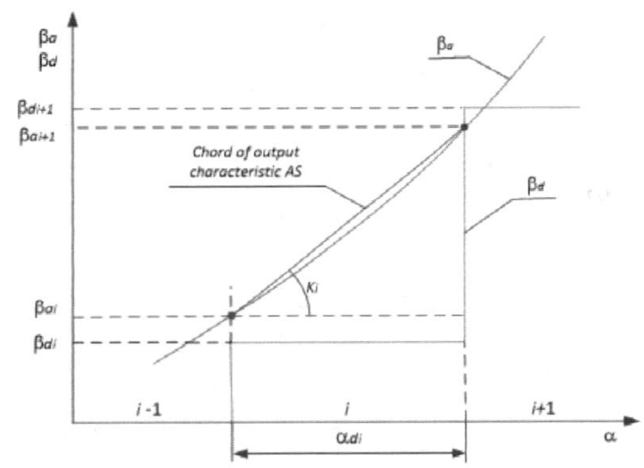

Figure 3. Calibration parameters of the i-th quants DS.

Note that the values of the parameters βa_i, βd_i and αd_i are determined during bench calibration. Performing rotation of the hybrid sensor shaft in the entire measurement range $\alpha\in[0, 2\pi]$, we obtain an array of the specified parameters for each i-th quants $(i = \overline{1, N})$. Also according to the formula (2) for each quants are calculated the values slope chord coefficient k_i of the output characteristic for analog sensor. The resulting arrays of parameters values βa_i, βd_i, αd_i and k_i are stored in the memory of the controller, and then they act as correction coefficients.

The calibration bench must have an external reference angle sensor. The reference sensor serves as a source of the angular position signal to determine the exact value of the angle at the moments of quants switching. This makes it possible to compensate for the instrumental error of the digital sensor, which can be caused by an error in the manufacture of the code disk or an error in the installation of the reading elements. Compensation is carried out by accurately measuring the parameters βd_i and αd_i during calibration, and taking these values into account in the calculation of the coefficients k_i values according to the formula (2), and then

IOP Conf. Series: Materials Science and Engineering **714** (2020) 012028 doi:10.1088/1757-899X/714/1/012028

in the working process when calculating the of value of angular position to the formula (1).

3. Results of mathematical modeling

Figures 4 and 5 present the results of mathematical modeling of the Hybrid Angular Position Sensor without correction and with correction, respectively.

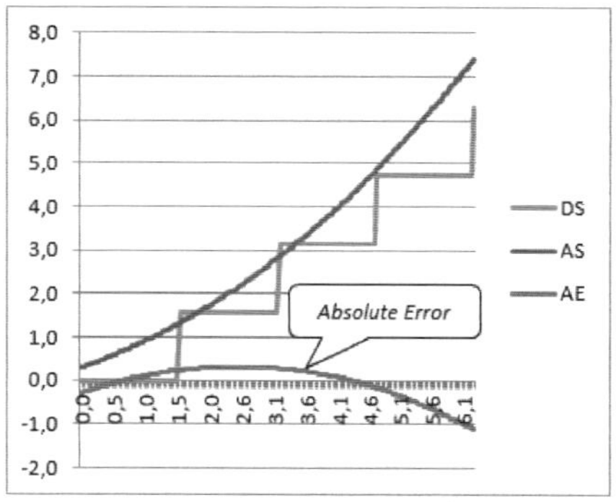

Figure 4. Without correction.

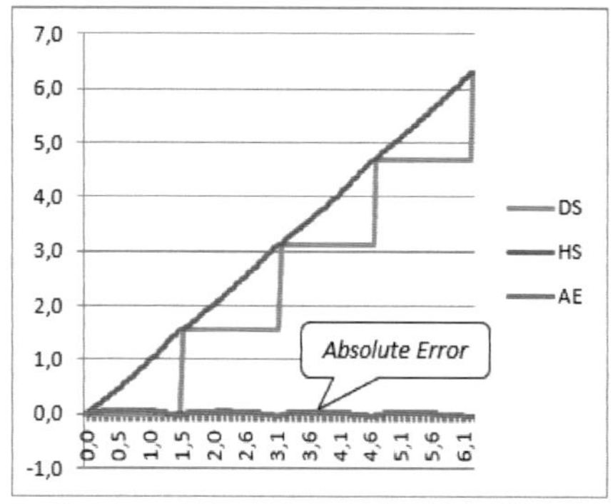

Figure 5. With correction.

In the process of modeling, the absolute error relative to the characteristics of the ideal sensor was calculated (figure 2) for the output characteristic of the analog sensor without correction, as well as the absolute error for the output characteristic of the hybrid sensor, i.e. with correction.

The analysis of the simulation results shows that if the output characteristic of the analog sensor is monotonic and has no discontinuities, then regardless of the degree of its curvature, the output characteristic of the hybrid sensor has a much smaller absolute error.

Further full-scale tests of the experimental sample fully confirmed the performance of the hybrid angular position sensor [3].

The proposed principle of hybridization can serve as a basis for angular position sensors of different designs, as well as can be used in sensors for determining the linear position.

4. Summary

Thus, Hybrid Sensors can significantly reduce the error of angular position measurement even when using relatively coarse Digital and Analog Sensors in their composition. This makes it possible to use sensors of smaller dimensions as part of the hybrid, and also reduce the environmental impact on the output characteristic of the sensor.

References

[1] Zayets V F, Kulabukhov V S, Tsatsin A A and Tuktarev N A 2019 Gibridnyy datchik izmereniya uglovogo polozheniya zayavitel' AO MNPK "Avionika" *Zayavka* no. 2019117876 ot 10.06.2019 Rossiyskaya Federatsiya, MPK H03M1 1/46

[2] Zayets V F, Kulabukhov V S, Tsatsin A A and Tuktarev N A 2019 Gibridnyy sposob izmereniya uglovogo polozheniya AO MNPK "Avionika" *Zayavka no*. 2019117875 ot 10.06.2019 Rossiyskaya Federatsiya, MPK H03M1 1/46

[3] Tsatsin A A, Perepelitsin A V, Orlov A I, Firsov A M and Kalym N P 2018 Naturnoye modeli-rovaniye gibridnogo datchika uglovogo polozheniya *Materialy V nauchno-prakticheskoy konferentsii pamyati O.V. Uspenskogo* pp 39–45

AUTHOR INDEX

Institute of Physics
Dirac House, Temple Back
Bristol BS1 6BE UK

ISSN: 1757-8981
ISBN 978-1-7138-2573-9

CURRAN ASSOCIATES INC.
proceedings.com